# The 10 Principles of Food Industry Sustainability

# The 10 Principles of Food Industry Sustainability

**Cheryl J. Baldwin**

*Pure Strategies, Gloucester, MA, USA*

WILEY Blackwell

*Registered Office*
John Wiley & Sons, Ltd., The Atrium, Southern Gate, Chichester, West Sussex, PO19 8SQ, UK

*Editorial Offices*
9600 Garsington Road, Oxford, OX4 2DQ, UK
The Atrium, Southern Gate, Chichester, West Sussex, PO19 8SQ, UK
111 River Street, Hoboken, NJ 07030-5774, USA

For details of our global editorial offices, for customer services and for information about how to apply for permission to reuse the copyright material in this book please see our website at www.wiley.com/wiley-blackwell.



*Library of Congress Cataloging-in-Publication Data*

Baldwin, Cheryl.
The 10 principles of food industry sustainability / Cheryl J. Baldwin.
pages cm
Includes bibliographical references and index.
ISBN 978-1-118-44773-4 (pbk.)
1. Food industry and trade. 2. Food supply. 3. Sustainable agriculture. 4. Farm produce. 5. Animal products. I. Title. II. Title: Ten principles of food industry sustainability.
HD9000.5.B3156 2015
664.0068′4–dc23

2014031889

A catalogue record for this book is available from the British Library.

Wiley also publishes its books in a variety of electronic formats. Some content that appears in print may not be available in electronic books.

Cover image: Food icons: ©iStock.com / missbobbit; Farm icons: ©iStock.com / missbobbit; Restaurant icons: ©iStock.com / Rakdee; Vector-ecology and Recycling icons: ©iStock.com / greyj; Globes icons and Symbols: ©iStock.com / filo; Business and Shopping icons: ©iStock.com / fairywong; Agriculture and Fisheries Color icons: ©iStock.com / godfather744431

Set in 10.5/13pt Times Ten by SPi Publisher Services, Pondicherry, India

1 2015

# Contents

# Acknowledgments

This book could not be put together without all the dedicated companies, professionals, and researchers in agriculture, food, nutrition, and related fields. It is because of the efforts of many that we have reached a point at which the leading principles of sustainability in the food industry can solidly emerge.

I would like to acknowledge the contributors of the book *Sustainability in the Food Industry* as the idea of the principles were derived from my effort as the editor of that book. I also want to acknowledge Green Seal's *Greening Food and Beverage Services* resource which I developed as pieces of that effort were used as a stating point, with permission from Green Seal, for this book.

Thank you to Tim Greiner and Pure Strategies for the valuable feedback on the draft concepts in the book and the opportunity to explore these issues as a consultant to companies advancing food sustainability.

Finally, my deepest appreciation goes to my husband and children, who were endlessly supportive and patient through the time that it took to put this book together. This book is dedicated to you.

# 1 Introduction to the Principles

## 1.1 The 10 principles of food industry sustainability

1. *Safe and highly nutritious food is accessible and affordable to promote and support a healthy population.*
2. *Agricultural production beneficially contributes to the environment while efficiently using natural resources and maintaining a healthy climate, land, water, and biodiversity.*
3. *Use of animals, fish, and seafood in the food supply optimizes their well-being and adds to environmental health.*
4. *Producer equity and rural economy and development are strengthened with fair and responsible production and sourcing.*
5. *Safe and suitable working conditions are provided to support employees across the supply chain.*
6. *Food and ingredient processing generates resources and requires minimal additional inputs and outputs.*
7. *Packaging effectively protects food and supports the environment without damage and waste.*
8. *Food and ingredient waste and loss are prevented across the supply chain and what cannot be avoided is put to a positive use.*
9. *Food and ingredients are efficiently delivered across the supply chain and to the consumer.*
10. *The supply chain and consumers advance sustainable business and food consumption.*

*The 10 Principles of Food Industry Sustainability*, First Edition. Cheryl J. Baldwin.

## 1.2 Principles–practices–potential

Our food system has the potential to produce renewable energy, replenish freshwater and other natural resources, provide an effective means of developing economic capacity, and remove waste through closing resource loops while nourishing the population. Are we achieving this potential? No, we fall short. In fact, we are not able to feed our population and yet cause astounding environmental and social damage.

The Principles of Food Industry Sustainability provide guidance on what to focus on across the supply chain to meet the needs of the population while not contributing to destruction of the environment or society. This book explains these principles through examples of how the supply chain has adopted them and what approaches are working, best practices. In many cases the efforts are moving past reducing detrimental impact and toward the goal of having meaningful and positive effects.

## 1.3 What is sustainability in the food industry?

The world's population depends on the food industry to produce, process, and deliver safe and nutritious food every day of the year. The demands on the food industry from farm to fork continue to increase. The global population is expected to increase from the current 7 billion to nearly 10 billion by 2050, thereby increasing food needs more than 60% (Consultative Group on International Agricultural Research [CGIAR] 2014). Most of the growth will be in developing countries where improvements in standards of living are rapidly shifting the diet from grains, beans, and other legumes to more animal protein. This will increase the demand for meat and poultry about 35% by 2015 (Pew 2008).

Yet, the food system is already contributing to widespread environmental damage and compromised health and livelihoods of our global population. The amount of energy used to produce, process, package, store, and transport food is seven and a half times the amount of energy the food actually provides in return (Heller & Keoleian 2000). The food supply is thereby a significant factor in climate change, water use and pollution, and the reduction of fish stocks in the oceans; at the same time 33% of adults in the United States are obese and over 12% of the world's population is malnourished.

The goal of sustainability in the food industry is to produce and consume food in a way that supports the well-being of generations. The current system clearly falls short and, with the growing demands for food as the population surges, there is the need and the opportunity for the food industry to balance the market needs for food with its environmental and social requirements. The ten principles given in this book provide a framework in which to address

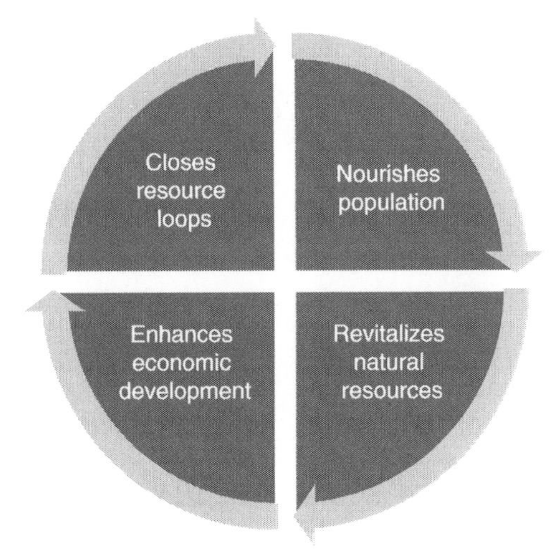


**Figure 1.1** Aim for a sustainable food system.

these requirements to move toward sustainability in the food industry. Together they aim for a food system that nourishes the population, revitalizes natural resources, enhances economic development, and closes resources loops (see Figure 1.1).

## 1.4 The destructive course of the food system

The many activities that go into the global food system can be divided into five major parts: agricultural production, processing and packaging, distribution and marketing, consumption, and waste (see Figure 1.2). Each stage of the food life cycle has unique interactions with the environment and society, causing problems that can be reduced or avoided. This book does not go into the details of these issues but provides a concise summary of the importance and relevance of addressing these concerns. The chapters that follow describe how the supply chain is moving away from the course of destruction toward a system of sustainability that has the potential to thrive economically and benefit both the environment and society.

The primary environmental and social issues of the current food system are closely related to each other and often influence each other. These include climate change, natural resource depletion and degradation, pollution and toxicity, rural economy and development, and food safety and nutrition (see Figure 1.3). These issues are not theoretical but are creating real challenges that businesses are facing today. One example of a supply chain disruption attributed to climate change was the unusually prolonged drought in Russia over the summer of 2010. By early August, more than one-fifth of Russia's wheat crop had been destroyed and the government banned all grain exports, contributing to wheat price futures reaching their highest point in nearly two

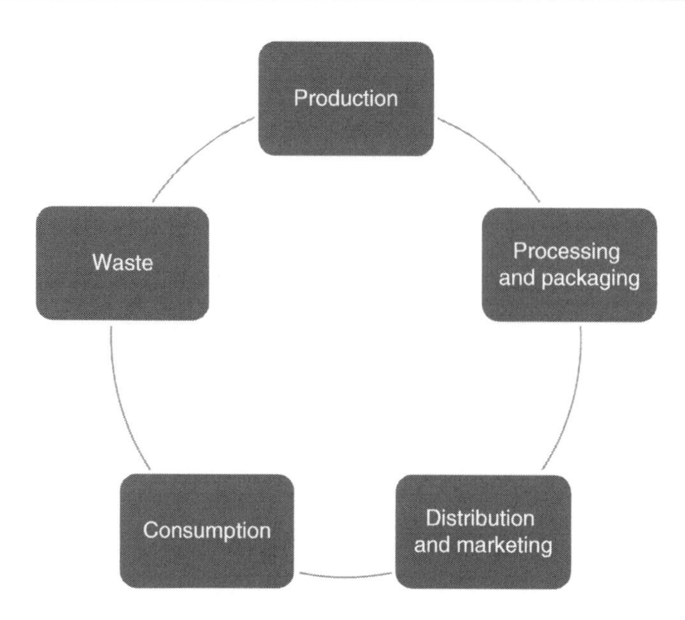


**Figure 1.2** Food life cycle.

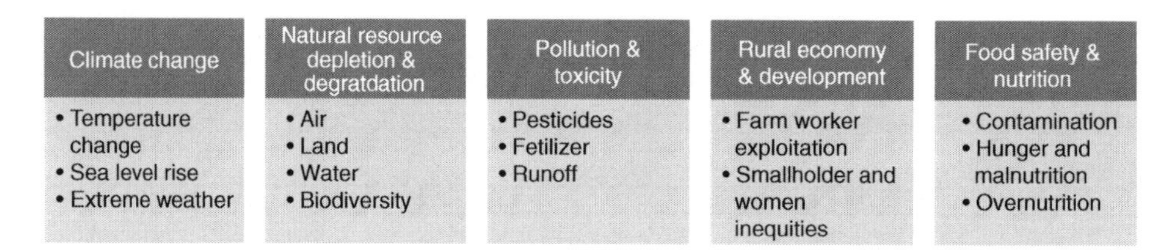


**Figure 1.3** Global environmental and social challenges with the food system.

years. General Mills was one of many food manufacturers that faced significant price pressure as a result and announced price increases of between 4 and 5% in September 2010 (Reed & Willis).

## 1.4.1 Climate change

Climate change is the shifting of global temperatures owing to various factors. A leading cause is the accumulation of heat-trapping gases in the atmosphere, called greenhouse gases (GHGs), including carbon dioxide, methane, nitrous oxide, and others. GHGs trap the sun's heat in the earth's atmosphere rather than allowing it to escape out to space. The accumulation and increase of GHGs has led to warmer atmospheric temperatures.

The food supply contributes significantly to climate change by being responsible for about 10 to 30% of global GHGs (U.S. Environmental Protection Agency [EPA] 2010a; and Bellarby et al. 2008). GHGs include the carbon dioxide emitted from electricity production and fuel use, methane from landfills and the production of an increasing number of livestock animals, nitrous oxide from excessive fertilizer use, and other kinds of emissions. Carbon dioxide is the most prevalent GHG; however, methane and nitrous oxide are significantly more potent than carbon dioxide at warming the climate (see Table 1.1). Table 1.2 factors in this difference in global warming potential (by putting all GHGs into equivalent terms) and shows how each of these GHGs contributes to the overall emissions from agriculture.

This abnormal rise in global temperature linked to GHGs has significant consequences. Current projections indicate that if GHG emissions are allowed to continue at their current pace, a temperature increase of 2 to 4.5 °C (3.6 to 8.1 °F) is likely by 2100 (United Nations [UN] 2010a). This temperature change is expected to increase the global sea level by 28 to 58 centimeters (11 to 23 inches) by the end of the 21st century, lead to a 20 to 30% extinction of species, and an increased frequency of heat stress, droughts, and flooding (Intergovernmental Panel on Climate Change [IPCC] 2007a). These climate change–related impacts will be experienced in different ways across the globe, including lost homes and land from rising waters and more human deaths caused by extreme weather and high temperatures. The temperature increases will modify growing seasons and shift where crops can be grown as well

**Table 1.1** Global warming potential (GWP) of a sample of GHGs (adapted from IPCC 2007b)

| GHG | GWP (carbon dioxide equivalent) |
|---|---|
| Carbon dioxide | 1 |
| Methane | 25 |
| Nitrous oxide | 298 |
| Hydrochlorofluorocarbons | 77–2310 |
| Chlorofluorocarbons | 4750–14400 |
| Hydrofluorocarbons | 124–14800 |

**Table 1.2** GHG emissions from food production (adapted from Bellarby et al. 2008)

| GHG and Its Source | Percentage of Total GHG Emissions (carbon dioxide equivalent) |
|---|---|
| Carbon dioxide from land conversion | 47.3 |
| Nitrous oxide from fertilized soils | 17.1 |
| Methane from enteric emissions | 14.4 |
| Methane and nitrous oxide from biomass burning | 5.4 |
| Methane from rice production | 5.0 |
| Carbon dioxide from irrigation and farm machinery | 4.2 |
| Carbon dioxide and nitrous oxide from fertilizer and pesticide production | 3.3 |
| Methane and nitrous oxide from manure | 3.3 |

as what crops are available. Weeds and pests will proliferate, thereby increasing the demand to further control insects, diseases, and weeds. Rainfall alterations are expected to increase the frequency of droughts and floods, decreasing yields and livestock productivity. Severe storms will potentially further damage crops. It is expected that there will be some benefits to certain crops, but the result overall will be less food available and at higher costs (International Food Policy Research Institute [IFPRI] 2009).

### 1.4.2 Natural resource depletion and degradation

Currently more than 60% of ecosystem services are being degraded or used faster than they can be replenished (World Resources Institute [WRI] 2005). These natural resources include clean air and water, uncontaminated soil, minerals, plants, fish, animals, and the life-supporting systems they sustain. Natural resources that took billions of years to produce are rapidly being lost. With the current rates of use and degradation, there may be few natural resources left by the end of this century (Hawken, Lovins, & Lovins 1999).

Topsoil is an important component to growing food and, thus, to supporting life on earth. However, topsoil is being lost at rates significantly greater than it is being formed. Agriculture, including overgrazing of livestock, is responsible for most of this loss. Adding to this, the productivity of the land is declining with agricultural practices including monocultures, overcultivation, and over-irrigation.

Similar to topsoil, freshwater is also a natural resource in decline. It is so vital to our function and well-being that in 2010 the UN General Assembly passed a resolution stating that access to safe and clean drinking water and sanitation is a human right, noting that 884 million people cannot obtain safe drinking water and that 1.5 million children under five years old die each year as a result of water- and sanitation-related diseases (UN 2010b). It is partially as a result of wasteful, unsustainable agricultural processes that freshwater is becoming a limited resource (Horrigan, Lawrence, & Walker 2002). Worldwide agriculture is responsible for two-thirds of freshwater use, largely by irrigation, but crops only use 45% of the water applied (FAO 1995). As a result, freshwater is becoming scarce, and quality freshwater even more scarce. These changes are occurring because of both the overuse of and the pollution of freshwater supplies.

Our food system has contributed significantly to biodiversity loss by intentionally reducing the variety of species used for food production, producing food in monocultures (i.e., a single crop grown on many acres of land), and destroying diverse habitats such as prairie and rainforest for agricultural production. This has resulted in agriculture being one of the primary sources of biodiversity loss (Convention on Biological Diversity [CBD] 2010). Biodiversity is the natural and vast array of different types of plants, animals, insects, and other life on earth. This array of species keeps a balance and order on earth; conversely, the loss of biodiversity puts us at risk. Such shifting can create devastating changes to ecosystems, including many unpredictable changes. What has been predicted, however, is that with the current rate of species losses (21 to 40% of species), plant growth will be reduced by 5 to 10% (Erickson 2012). This impact is comparable to what is expected from climate change, and yet this sizable risk is often overlooked in the discussion (Erickson 2012). For example, conversion of tropical forests to palm oil plantations has caused a loss of 73 to 83% of the bird and butterfly species (CBD 2010). Loss of pollinators such as these is already threatening food production, which relies on birds and insects for pollination. Biodiversity also helps nature cope with climate change so its loss will exacerbate the impact of global warming (Diaz, Tilman, & Fargione 2005). This may be just the tip of the iceberg as there are many consequences of biodiversity loss that are not fully understood.

### 1.4.3 Pollution and toxicity

The use of materials that are toxic to humans, animals, and the environment are widespread and come from packaging, agricultural inputs (fertilizer, pesticides, herbicides), fuels, and cleaning products. Toxic materials present both short-term and long-term threats. For example, the long-term effects of pesticides include cancer and disruption of the body's reproductive, immune, endocrine, and nervous systems (Horrigan et al. 2002). The use of such toxic chemicals has had mixed results and yet the application of pesticides is increasing (Malakof and Stokstad 2013, Hawken et al. 1999). The long-term effects of nitrogen fertilizer are also deadly. Plants absorb one-third to one-half of the nitrogen fertilizer applied, which means that at least half of the fertilizer finds its way into the soil and waterways as toxic runoff. Aquatic plants grow faster when fertilized by such nutrient-rich run-off. When they die, extra oxygen is required for their decomposition. If this process continues long enough, areas of the ocean suffer depleted oxygen stores. This results in dead zones that cannot support fish or any other marine life. Excessive use of chemical fertilizer is also detrimental to the soil by making it more acidic or less fertile (Horrigan et al. 2002). Adding to this, agriculture is estimated to be responsible for 70% of the pollution of rivers and streams (EPA 1998). With freshwater supplies along with other natural resources declining, this pollution compounds other problems.

### 1.4.4 Rural economy and development

The global food demand is estimated to increase 60% by 2050 (CGIAR 2014). To meet this demand more production will be needed from developing countries and small-scale producers, who already supply about 80% of the food (Fan 2011). However, these producers are the poorest and have little resources to make farming improvements. To add to this, in many regions women contribute the bulk of farm labor but are not allowed access to the same resources as men. Women receive only about 1% of all agricultural financing or credit (Fraser 2009). This inequity compromises their agricultural productivity and earning potential. Agricultural jobs also are characterized by high risks and low wages. In the United States, the fatality rate for farm workers is seven times higher than that of other private industry jobs (McCluskey, McGarity, & Shapiro 2013). Even today there remain instances of agricultural laborers being exploited, enslaved, and abused. Our food system cannot survive if those on whom we rely on to grow the food are so unfairly treated.

### 1.4.5 Food safety and nutrition

The fundamental purpose of our food supply is to provide safe and nutritious food for the population. Despite notable advances in food safety, contamination with pathogens and chemicals remains a concern. Even in developed countries such as the United States food-borne illness is widespread, causing nearly 50 million illnesses and 3,000 deaths annually (U.S. Centers for Disease Control and Prevention [CDC] 2012). The nutritional status of the population needs to be improved. Globally nearly a billion people are hungry, over a billion are overweight, and another half a billion are obese (FAO 2012). These health conditions compromise the ability of our population to prosper. And yet, much of this unbalance is preventable as enough food is currently produced to feed the population.

## 1.5 Reasons for principles for sustainability in the food industry

The Principles for Sustainability outlined in this book bring together the leading issues that need to be addressed to advance toward sustainability. Too often there is a focus on just a few areas of concern, such as a limited set of environmental issues. Although working on priorities is critical for companies to get started, material issues can get overlooked, or worse, there may be unintended consequences or burden-shifting because there was not a holistic view of the issues. Sustainability requires a broader perspective: The Principles provide this view. They concisely outline each of the leading issues that, if advanced, would take us closer to sustainability in the food supply.

The Principles were intentionally positioned with a positive perspective. Instead of stating what should not be done, they articulate what needs to be done. This provides motivation to see what is possible and to aim beyond doing less harm and rather move to delivering an overall benefit.

The Principles outline the leading hotspots across the supply chain that address climate change, natural resource depletion, pollution and toxicity, rural economy and development, and hunger and nutrition. The issues are so widely applicable to the food supply that the Strategy for Sustainable Farming and Food in the United Kingdom and other organizations have developed a similar set of key principles for a sustainable food chain (Department for Environment, Food and Rural Affairs [DEFRA] 2006).

The Principles are based on the body of literature available on food, environmental, and social issues. This includes life-cycle assessment, hotspot analysis, work by companies and the industry, and other resources. Discussion of economic issues is generally limited to social development in this book because other financial considerations are well understood by companies managing their risks and costs every day – however, economic sustainability of businesses is important to be able to deliver on the other needs and as a result it is included in the principles. This book fills the need of providing a practical view of how to bring environmental and social considerations into the supply chain through business actors. In many cases the actions are more heavily weighted to companies further down the supply chain, such as manufacturers and retailers, but there are roles for each stage of the supply chain.

The Principles outline the basic needs for the food supply to be safe and nutritious—this is the fundamental purpose of the system. Meeting this need requires improvements in farming and downstream activities, including manufacturing, distribution, food service, and retailing, such that they support a healthy environment and demand fewer resources and even generate resources such as energy and water. Animal-based production systems and harvesting of fish and seafood have unique considerations because these products have significantly greater life-cycle impacts than other products and the treatment of the animals is critical to the effectiveness and sustainability of the system.

To ensure the long-term food supply and provide for needed economic advancement, diversification of sourcing with development support for growers and suppliers should be practiced to provide fair-market access for producers, especially for small-scale operators and women in agriculture. It is also important to address worker safety and treatment at the farm and throughout the supply chain.

There is far too much waste in the supply chain. Food waste is particularly egregious with about 30% of what is produced being wasted (Gustavsson, Cederberg, Sonesson, van Otterdijk, & Meybeck 2011). This futile use of resources needs to be fixed to help reduce the demand for more production and feed the population. Packaging waste is a unique challenge as it provides an important function of protecting the food. Packaging already has been a consistent area of focus for sustainability initiatives, but there is significant room for improvement to minimize material impacts and eliminate waste. More emphasis is needed to ensure that packaging waste has a useful purpose, such as through recycling, to close the resource loop and make the most out of the material.

The Principles end with the role of the consumer and the opportunity the supply chain has to ensure its economic viability and to assist the consumer in moving to sustainable consumption and supporting each of the other Principles to nourish the population, revitalize natural resources, enhance economic development, and close resources loops.

## 1.6 The business benefit

The concept of sustainability includes economic considerations along with environmental and social ones. There are business benefits for effectively engaging in the environmental and social issues. The World Business Council for Sustainable Development has found that businesses that incorporate sustainable practices benefit by increased financial success, including more resilience. Organizations committed to sustainability financially outperformed industry averages by 15% during the economic downturn of 2008 to 2009 (Mahler, Barker, Belsand, & Schultz 2009). This is because sustainability helps reduce internal costs of operating, address reputational risks and risks in the supply chain that can threaten the business, meet customer and consumer demands to enable growth, and improve organizational effectiveness and business relationships (see Figure 1.4). For example, companies that have supply-chain disruptions experience an average share-price decline of nearly 20%, higher share-price volatility, and lower return on sales and return on assets (International Finance Corporation [IFC]). Sustainability efforts can alleviate some of these risks. Cost savings are a common benefit, such as a 1 to 3% savings being typical with facility operations improvements through sustainability programs (Strandberg 2009). Additionally, sustainability enhances employee productivity by about 2% and has become a critical factor for employees when considering what company to work for (Industry Canada).

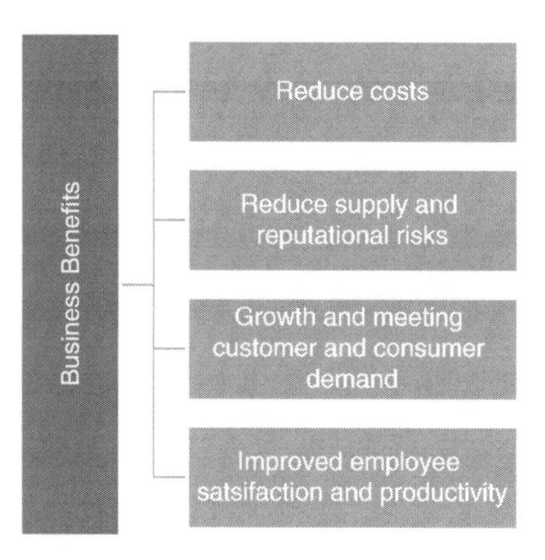


**Figure 1.4** Business benefits of sustainability.

Multiple business benefits are to be gained by integrating sustainability into company efforts. Some of the benefits have a clear payoff, such as in energy conservation and reduced utility bills. Other opportunities may present longer-term value, such as building farmer capabilities and a more resilient supply. The business cases discussed through the chapters are not detailed with specific dollar amounts to instead bring the focus to the best practices that companies are leveraging to realize the combination of intended benefits: economic, environmental, and social.

## 1.7 What needs to be done

We can no longer rely on business as usual. Companies need to adopt best practices and explore innovations across the supply chain to begin to realize not just a more sustainable food supply that feeds the population but also one that provides an overall benefit to the economy, environment, and society.

If this vision is not compelling enough to see the need for change, there are several drivers pushing companies to take action (see Figure 1.5), including emerging regulations, customer demands, and escalating costs and risks. Remember, the business benefits are very real and are a big motivator. In addition to industry action, there is a need for significant adjustments to governmental policy and consumer behavior. These topics are beyond the scope of this book's discussion but are important for reaching the vision.

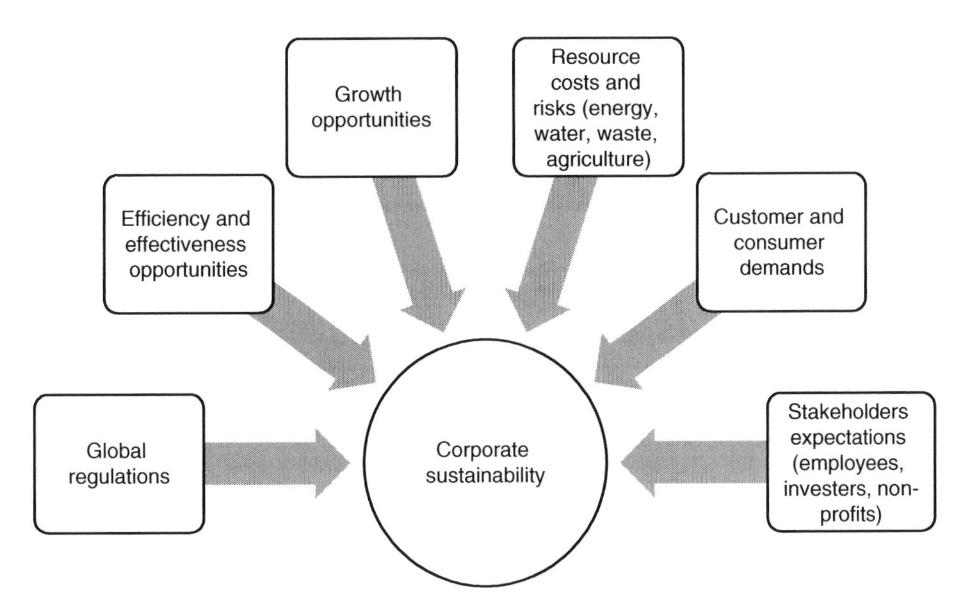


**Figure 1.5** Business drivers of sustainability.

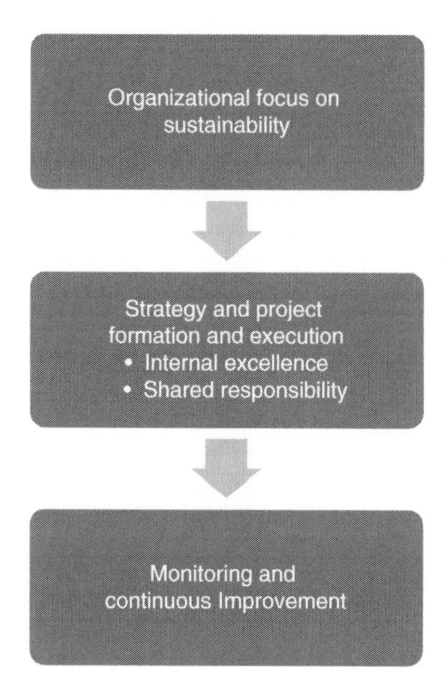


**Figure 1.6** Business sustainability approach.

Companies looking to get started begin by committing to a focus on sustainability, set appropriate goals, assess needs and priorities, and address opportunities by incorporating sustainability into business practices and tracking progress (see Figure 1.6). Every one of the Principles cannot be worked on right away nor all at once. The best way to begin is to select a few areas of focus and achieve some successes before taking a broader approach (i.e., crawl before you walk and walk before you run). Do this with a view to all of the issues and Principles to ensure problems are not shifting from one place to another. Companies will often start with an internal focus on operational efficiencies or start with pilot projects. Then after their initial approach, they realize some success and more opportunities are identified, and the effort typically expands. Part of the evolution includes addressing areas of shared responsibility and value. As more companies reach beyond their four walls and work collaboratively through the supply chain, exciting progress and innovation can occur for shared value. Many of the best practices described in this book are illustrated with examples from companies engaged in advancing sustainability both internally and externally. Some of the examples, however, include companies just getting started; they demonstrate that every step forward is an important part of the journey as long as they keep the destination is in mind—nourishing the population, revitalizing natural resources, enhancing economic development, and closing loops in resources.

# References

Bellarby, J., Foereid, B., Hastings, A., Smith, P. et al. (2008). *Cool farming: Climate impacts of agriculture and mitigation potential.* Retrieved from http://www.greenpeace.org/international/Global/international/planet-2/report/2008/1/cool-farming-full-report.pdf

Consultative Group on International Agricultural Research (CGIAR). (2014). *Undernourishment and obesity*. Retrieved from http://ccafs.cgiar.org/bigfacts2014/#theme=food-security&subtheme=undernourishment

Convention on Biological Diversity (CBD). (2010). *Global biodiversity outlook 3.* Retrieved from http://www.cbd.int/doc/publications/gbo/gbo3-final-en.pdf

Department for Environment, Food and Rural Affairs (DEFRA). (2006). *Food industry sustainability strategy*. Retrieved from https://www.gov.uk/government/publications/food-industry-sustainability-strategy-fiss

Diaz, S., D. Tilman, D., & Fargione, J. (2005). Biodiversity regulation of ecosystem services. In *Ecosystems and human well-being: Current state and trends*. Washington, DC: Island Press.

Erickson, J. (2012). *Ecosystems effects of biodiversity loss could rival impacts of climate change, pollution*. Retrieved from http://ns.umich.edu/new/multimedia/slideshows/20366-ecosystem-effects-of-biodiversity-loss-could-rival-impacts-of-climate-change-pollution

Fan, S. (2011). *Leveraging smallholder agriculture for development*. Retrieved from http://www.slideshare.net/shenggenfan/leveraging-smallholder-agriculture-for-development

Food and Agriculture Organization of the United Nations (FAO). (2012). *Improving food systems for sustainable diets in a green economy*. Retrieved from http://www.fao.org/fileadmin/templates/ags/docs/SFCP/WorkingPaper4.pdf

Food and Agriculture Organization of the United Nations (FAO). (1995). *Water: a finite resource in dimensions of need: An atlas of food and agriculture*. Retrieved from http://www.fao.org/docrep/u8480e/U8480E0c.htm#Irrigation

Fraser, A. (2009). Harnessing agriculture for development. Retrieved from http://www.oxfam.org/sites/www.oxfam.org/files/bp-harnessing-agriculture-250909.pdf

Gustavsson, J., Cederberg, C., Sonesson, U., van Otterdijk, R., & Meybeck, A. (2011). *Global food losses and food waste: Extent, causes and prevention* (from the Food and Agriculture Organization of the United Nations). Retrieved from http://www.fao.org/docrep/014/mb060e/mb060e00.pdf

Hawken, P., Lovins, A., & Lovins, L. H. (1999). *Natural capitalism: Creating the next industrial revolution*. New York: Back Bay Books/Little, Brown and Company.

Heller, M. C. & Keoleian, G. A. (2000). *Life cycle-based sustainability indicators for assessment of the US food system.* Center for Sustainable Systems, University of Michigan, Report CSS00-04. Retrieved from http://css.snre.umich.edu/css_doc/CSS00-04.pdf

Horrigan, L., Lawrence, R., & Walker, P. (2002). How sustainable agriculture can address the environmental and human health harms of industrial agriculture. *Environmental Health Perspectives,* 110, 445–456.

Industry Canada. *The business case*. Retrieved from http://www.ic.gc.ca/eic/site/csr-rse.nsf/eng/rs00555.html

Intergovernmental Panel on Climate Change (IPCC). (2007a). *IPCC fourth assessment report: Climate change 2007*. Retrieved from http://www.ipcc.ch/publications_and_data/ar4/wg2/en/contents.html.

Intergovernmental Panel on Climate Change (IPCC). (2007b). *IPCC fourth assessment report: Climate change 2007*. Retrieved from http://www.ipcc.ch/publications_and_data/ar4/wg1/en/ch2s2-10-2.html.

International Finance Corporation (IFC). *The business case for sustainability*. Retrieved from http://www.ifc.org/wps/wcm/connect/9519a5004c1bc60eb534bd79803d5464/Business%2BCase%2Bfor%2BSustainability.pdf?MOD=AJPERES

International Food Policy Research Institute (IFPRI). (2009). *Climate change: Impact on agriculture and costs of adaptation*. Retrieved from http://www.ifpri.org/sites/default/files/publications/pr21.pdf

Mahler, D., Barker, J. ,Belsand, L., & Schultz, O. (2009) *Green winners: The performance of sustainability-focused companies during the financial crisis*. A.T. Kearny Publication. Retrieved from http://www.atkearney.com/images/global/pdf/Green_winners.pdf

Malakof, D., & Stokstad, E. (2013) Pesticide planet. *Science*, 341 (6147), 730–731. DOI: 10.1126/science.341.6147.730

McCluskey, M., McGarity, T., & Shapiro, S. (2013). *At the company's mercy: Protecting contingent workers from unsafe working conditions*. Retrieved from http://www.progressivereform.org/articles/Contingent_Workers_1301.pdf

Pew Charitable Trusts. (2008). *Putting meat on the table: Industrial farm animal production in America*. Retrieved from http://www.pewtrusts.org/our_work_report_detail.aspx?id=38442

Reed, D., & Willis, C. *Sustaining the supply chain*. Retrieved from http://www.pwc.com/gx/en/governance-risk-compliance-consulting-services/resilience/publications/sustainable-supply-chain.jhtml

Strandberg Consulting. (2009). *The business case for sustainability*. Retrieved from http://www.corostrandberg.com/pdfs/Business_Case_for_Sustainability_21.pdf

United Nations (UN). (2010a). *Gateway to the UN system's work on climate change: The science*. Retrieved from http://www.un.org/wcm/content/site/climatechange/pages/gateway/the-science

United Nations (UN). (2010b). *General assembly adopts resolution recognizing access to clean water, sanitation as human right, by recorded vote of 122 in favor, non against, 41 abstentions press release July 28, 2010*. Retrieved from http://www.un.org/News/Press/docs/2010/ga10967.doc.htm

U.S. Centers for Disease Control and Prevention (CDC). (2012). *Foodborne illness, foodborne disease, (sometimes called "food poisoning")*. Retrieved from http://www.cdc.gov/foodsafety/facts.html

U.S. Environmental Protection Agency (EPA). (2010). *Emissions inventory 2010. Inventory of US greenhouse gas emissions and sinks: 1990–2008*. Retrieved from http://www.epa.gov/climatechange/emissions/usinventoryreport.html

U.S. Environmental Protection Agency (EPA). (1998). *Reducing water pollution from animal feeding operations: Testimony before subcommittee on forestry, resource conservation, and research of the committee on agriculture, U.S. House of Representatives, 13 May 1998*. Retrieved from http://www.epa.gov/ocirpage/hearings/testimony/105_1997_1998/051398.htm

World Resources Institute (WRI). (2005). *Millennium ecosystems assessment, 2005: Ecosystems and human well-being synthesis*. Retrieved from http://www.maweb.org/documents/document.356.aspx.pdf

# 2 Agriculture and the Environment

*Principle: Agricultural production beneficially contributes to the environment while efficiently using natural resources and maintaining a healthy climate, land, water, and biodiversity.*

Agricultural production is the dominant source of environmental degradation coming from the food supply. Climate, land, biodiversity, water, and pollution impacts have led to short- and long-term food production concerns. Because these environmental concerns come with notable risks and costs, efforts to address them are critical and have already begun. Even companies that do not directly interact with farmers are becoming more involved upstream. With this collaborative approach to addressing the supply chain impacts, progress is beginning to surge forward, such that agricultural production can serve the positive land and resource stewardship role it has the potential to be.

## 2.1 Climate

The food supply contributes significantly to climate change, responsible for up to one-third of global greenhouse gasses (GHGs) (Bellarby, Foereid, Hastings, & Smith 2008). This occurs through the release of carbon dioxide from deforestation and fossil fuel energy use, methane from livestock and rice production, and nitrous oxide from fertilizers. GHGs absorb the heat from the sun and trap it in the atmosphere rather than allowing it to pass through and out into space. As GHGs increase, the trapped heat increases, causing global warming and related climate change effects. Temperature and sea level increases result with a range of consequences from more frequent extreme weather events (e.g., heat stress, drought, and flooding) to extinction of species and other environmental damage.

*The 10 Principles of Food Industry Sustainability*, First Edition. Cheryl J. Baldwin.

Food production is vulnerable to these changes. The temperature increases will modify growing seasons and shift where crops can be grown and what crops are available. Weeds and pests will proliferate, increasing the demands to control insects, diseases, and weeds. Rainfall alterations are expected to result in increased frequency of droughts and floods, decreasing yields. This has already been experienced with record heat and drought levels reached in the United States in 2012, impacting 60% of crop production and close to 90% of corn and soybean crop production (Nierenberg & Reynolds 2012). Severe storms will further increase potential damage to crops. It is expected that there will be some benefits for certain crops but overall there will be less food available and at higher costs (International Food Policy Research Institute [IFPRI] 2009).

Carbon dioxide is the single largest GHG from agriculture. Carbon dioxide emissions come primarily from the conversion of land for agriculture (Bellarby et al. 2008). This is because the natural habitat of forests, grasslands, and wetlands that are destroyed to make room for agriculture are more effective at absorbing and storing carbon dioxide (Bellarby et al. 2008). When land is converted to agriculture the balance of carbon storage is shifted and results in more GHGs in the atmosphere. There are techniques that can reduce the total amount of carbon dioxide emitted from agricultural land. Proven practices to help with this include using cover crops to protect and nourish the soil, reducing tilling of the soil (e.g., no-till production), well-managed organic fertilization, and proper management of grazing land (Bellarby et al. 2008). These techniques also help retain soil health for the land to be productive longer, reducing the need to convert new land to farmland (Picone & Van Tassel 2002).

Agricultural production also emits carbon dioxide through the use of energy and fossil fuels to run machinery and irrigate the land. The production of inputs such as fertilizer and pesticides is energy-intensive and contributes to carbon dioxide emissions. There are production methods that use less of these synthesized fertilizers and pesticides, such as integrated pest management and conservation agriculture. Growing produce in heated greenhouses, or hot houses, comes with a high fossil fuel energy demand that can be avoided with alternative energy sources or with passive heat (sunlight).

Another important GHG in agriculture is methane. This GHG is 25 times more potent than carbon dioxide in its global warming potential (IPCC 2007). The main source of methane is livestock production, contributing about 60% to global methane emissions (Bellarby et al. 2008). There is more discussion about livestock production in the next chapter. Methane is also generated in rice production. Rice is commonly produced in swampy, flooded conditions that promote the growth of microorganisms that produce methane. There are different rice varieties that do not need the level of moisture that is prone to microbial production of methane and there are other mitigation approaches to reducing methane production from rice production.

The other key GHG in agriculture is nitrous oxide. Nitrous oxide is 298 times more powerful than carbon dioxide in global warming potential (IPCC 2007). Synthetic fertilizer is responsible for more than 75% of nitrous oxide emissions in food production (Lappé 2010). This is primarily because fertilizer is overused and not taken up well by the plants; so some of the surplus is released into the atmosphere (Bellarby et al. 2008). Approximately 25 to 30% of applied nitrogen from fertilizer is lost as nitrous oxide emissions (Galloway et al. 2004; FAO 2012). This can be avoided with best practices such as using biological nitrogen fixation with legumes which has been estimated to reduce GHG emissions by half (Drinkwater & Snapp 2007). Approaches such as nitrogen fixation are leveraged in organic production yielding wheat and corn with significantly lower GHG emissions than conventional production. The combination of such nutrient management approaches can dramatically reduce overall nitrous oxide emissions.

GHG emissions can be reduced and removed from the atmosphere through widespread adoption of existing agricultural best practices. These include building soil fertility with minimizing tillage, cover cropping, using diverse production systems such as agroforestry, and others (Nierenberg & Reynolds 2012). Preserving biodiversity used for food production and conserving water also are important (Nierenberg & Reynolds 2012). These approaches reduce the need for land conversion, chemical inputs, and irrigation (Nierenberg & Reynolds 2012). The result is that close to 90% of the GHG emissions in agriculture could be avoided or mitigated and there is potential that agriculture can lead to serving a beneficial role in addressing climate change (Reynolds & Nierenberg 2012).

## 2.2 Land and biodiversity

Fertile land is critical for food production, but some agricultural practices have led to rapid declines in soil fertility and available land for production. The amount of arable land is 60% of what it was 50 years ago (Horrigan, Lawrence, & Walker 2002). Poor farming practices are responsible for most of this loss, damaging about 38% of all farmland (Horrigan et al. 2002). Soil erosion, desertification, and inefficient use of the land are significant contributors to land loss. There are limits in the amount of land expansion possible. In order to meet the needs of the growing population, the existing land needs to be nurtured (FAO 2012).

Topsoil is being lost at rates significantly greater than it is being formed (World Economic Forum [WEF] 2012). The main considerations for soil degradation are the degree of soil coverage by vegetation, the intensity and frequency of tillage, the use of heavy machinery and grazing, fertilization and nutrient management, and the application of chemicals (FAO 2012).

A growing amount of land is drying out and degrading to the point that it is no longer productive. This process of desertification is caused by many factors, including poor agricultural practices such as over cultivation, overgrazing, and overuse of water. It is also caused indirectly when land is deforested to create new cropland or new pastures for livestock. As a result, almost 15% of all land surfaces may already be experiencing some degree of desertification (Horrigan et al. 2002).

Agriculture is also one of the primary sources of biodiversity loss (Convention on Biological Diversity [CBD] 2010). This is because food production impacts each of the main drivers for biodiversity loss: habitat change/destruction, overexploitation of natural resources, climate change and emissions, pollution, and invasive species (FAO 2012). Biodiversity provides resilience to life on earth. Loss of species means a risk of life on earth going out of balance, resulting in unpredictable changes. Some changes that have been seen include irreversible land damage (erosion, desertification), uncontrollable invasion of weeds or other pests, and complete devastation of plants or crops from disease. The cost and long-term impacts of biodiversity loss is not fully understood, but the available estimates highlight its importance.

The natural services provided by biodiversity are critical to food production. The pest control mechanism served through natural biodiversity is estimated to save $13.6 billion per year in the United States (Power 2010). About 40% of the food crops are pollinated from animals including bees and birds (Power 2010). Honeybees have been experiencing dramatic losses from a combination of causes including pesticides, declining food sources, and disease (USDA 2012). The FAO estimated that insect pollination services are worth $200 billion and if lost would significantly disrupt fruit, vegetable, cocoa, tea, and coffee production and result in nutrient deficiencies in global diets (Gallai, Salles, Settele, & Vaissiere 2009).

Monoculture production illustrates potential issues with biodiversity decline. Producing food in monocultures (the production of a single crop across large plots of land) leads to reduced resistance to drought and other challenges. Pests such as insects and pathogens (disease-causing organisms) can find their food sources more easily in monocultures than in diverse crop mixtures. Monocultures also have lower populations of the natural enemies of pests such as spiders, wasps, dragonflies, and predatory beetles. As a result, the use of monocultures increases a farm's dependence on pesticides and these pesticides also kill the natural enemies of pests (Picone & Van Tassel 2002). This is further complicated because pest species evolve resistance to pesticides much faster than the natural enemies of those pests, thus pest populations quickly recover (Picone & Van Tassel 2002). The loss of natural enemies from monocultures and pesticides may result in new problems such as "secondary pests"—those initially not an issue but become one due to the loss of natural enemies. The result of pest resistance and secondary pest outbreaks has been increasing the amounts of pesticides being applied which may lead to more toxic chemicals being used. Crop loss from pests has not decreased despite an increase in the use of pesticides over the last forty years (Oerke 2006). A 13% crop loss occurred in 1989 whereas in 1945 there was a 7% loss with a lower reliance on crop protection chemicals (Picone & Van Tassel 2002). In addition, pesticides kill wild bees and other beneficial species (Horrigan et al. 2002).

Biodiversity loss is a natural process, however, over the past 100 years humans have accelerated the extinction rate by at least 100-1000 times the natural rate (FAO 2006). This includes the rich genetics within a species and the total number of species from animals, plants, and microbes, and ecosystem diversity (FAO 2006). More than 90% of crop varieties have disappeared from farmers' fields with the majority of the world's population fed by less than 20 plant species and 14 animal species (FAO 2006). Just two apple varieties account for more than 50% of the entire crop in the United States and 99% of all turkeys in the United States are from one breed (Shand 2000). This loss comes from intentionally reducing the variety of species used for food production. The situation is expected to become more complicated with the expansion of genetic engineering, the use of laboratory techniques to transfer specific genes from one species to another. Genetic engineering (also referred to as genetic modification) has been used to add select traits such as herbicide and pest resistance. However, the potential of these benefits lies in careful application in practice to ensure that it doesn't lead to weed and pest resistance or detrimental decline in biodiversity, nor other concerns (Cox & Jackson 2002).

Biodiversity also provides the nutritional variety we need for a healthy diet. This is so critical that biodiversity is classified as a "fundamental determinant of health" by the World Health Organization (WHO). Genetic variations across and within crops provide important differences in nutritional composition such as in fruit, grains, vegetables, meat, and dairy products. For example, the protein, amino acid, B vitamin, fatty acid, and vitamin E content vary for different wheat crops (Institute of Medicine 2014).

Land conversion for agriculture is one of the key drivers for biodiversity loss because of its vast coverage of 30 to 40% of all land mass (CBD 2010). Livestock production is a leading source of biodiversity loss due to its increased requirement for land conversion for grazing and feed production (FAO 2006). Crop production also contributes to deforestation. Palm oil is the world's most consumed vegetable oil and is projected to increase 65% by 2020 (World Wildlife Fund [WWF] 2012). Palm production has come with significant deforestation. About one-third of deforestation in Indonesia and Malaysia between 2005 and 2010 was from land conversion to palm plantations (WWF 2012). Soybean and sugar cane production have similar habitat destruction issues (WWF 2012). Such habitat destruction contributes to biodiversity loss, climate change, and other damage.

Protecting biodiversity helps protect soil fertility and land productivity. Land conversion can be avoided by nurturing the fertility of the existing land.

Best practices to meet these multiple needs include the incorporation of organic matter in the soil, crop rotation, cover crops, managed irrigation, limited fertilization, minimum tillage, soil cover (mulching, managed fallow), and others (FAO 2012). If land conversion is necessary, then production techniques should support biodiversity and portions of the land should be set aside and protected. The practices that help support biodiversity include growth of diverse species (instead of monocultures), organic production and integrated pest management, well-managed grazing, and integrating habitat within production acreage (e.g., agroforestry, in which animals graze in forested areas).

## 2.3 Water and pollution

Access to clean water is essential for human life as well as for food production. The reality is that significant parts of the world do not have enough water. This is projected to worsen with over 60% of the world's population expected to live in water-stressed areas by 2025 (FAO 2006). Agriculture is both dependent on water and straining the availability of water. Agriculture is responsible for two-thirds of fresh water use and 70% of the pollution of rivers and streams (FAO 1995, U.S. Environmental Protection Agency [EPA] 1998). This pollution comes from excess nutrients, livestock production, and pesticides.

The agricultural sector is the largest user of water worldwide, consuming more than twice that of industry (23%) and dwarfing municipal use (8%) (WWF a). Yet, only 45% of the irrigation water is effectively used (FAO 1995). The FAO found that, "most irrigation systems across the world perform below their capacity and are not adapted to the needs of today's agriculture" (FAO 2011). This wasteful use of water for irrigation reduces the amount of water that is available and suitable for other uses (Horrigan et al. 2002). Furthermore, irrigation uses natural reservoirs of water including fossil water sources that once used up are gone forever. This loss has reduced the availability of food in some parts of the world. The High Plains Aquifer in the United States has been drained through irrigation to the point that parts of Texas and Kansas no longer have enough water available for farmer's needs (Wines 2013).

Beyond water use, irrigation uses energy and fossil fuels, further adding to the overall environmental damage. Excessive irrigation can lead to increased soil salinity and degraded water quality. Soil salinization occurs when irrigation water leaves behind salts. The salt content in the irrigated land then increases beyond the ability for the soil to be productive. Salinization has caused the abandonment of ten million hectares each year worldwide and affects 28% of the irrigated land in the United States (Pimentel et al. 2008, Horrigan et al. 2002).

Managing soil health, diversifying crop production, and reducing of livestock production can decrease water demands for agriculture. Conservation tillage or mulching, for example, can reduce soil's water evaporation by 35 to

50% (Power 2010). Water use can be further conserved by monitoring soil and plant water needs to time irrigation effectively and with efficient irrigation methods including some drip and sprinkler irrigation technologies. These and other approaches to conserving water not only protect this natural resource, they also contribute to increasing crop production (Rost et al. 2009).

Water impacts can be measured as a "water footprint." Food consumption is the largest contribution to the water footprint of an adult in the United States (National Geographic 2010). This is because it takes more water to produce food than other human activities. Energy production requires about 0.15 gallons of water for each kilowatt-hour from coal where it takes 119 gallons of water to produce a pound of potatoes and 880 gallons of water to produce a gallon of milk (National Geographic 2011). This is in part because water is needed to grow plants and animals, but a water footprint measurement also accounts for the amount of water polluted which is an important consideration for food production.

Agriculture is one of the leading causes of pollution. Pesticides, fertilizers, and other toxic farm chemicals pollute fresh water supplies, marine ecosystems, air, and soil. The EPA estimates that 70% of river and stream pollution is caused by agriculture from chemicals (pesticides, herbicides, and fertilizers), silt, and animal waste (EPA 1998). When these toxins get into the environment, they lead to harmful effects such as illness and death to humans and wildlife.

Nitrogen and phosphorus are commonly added to crops as fertilizer to improve productivity. The uptake of these additives in the field is limited and any unused/unabsorbed nutrients end up in the environment. While food production uses about 80% of the mined phosphorus, only 20% contributes to the growth of the crop with the rest ending up in the environment (FAO 2012). Approximately 50% of applied nitrogen ends up used by the plants with 20% of applied nitrogen ending up in aquatic ecosystems (the rest being emitted into the atmosphere) (FAO 2012, Power 2010). These excess nutrients cause harm to the environment and human health. A widespread environmental problem is dead zones where the nutrient load creates an artificial surge in aquatic plant and algae growth, followed by mass death and decomposition that depletes the available oxygen for other aquatic life - called eutrophication (Picone & Van Tassel 2002). This effect has created forty large, oxygen-starved "dead zones" around the world's waterways, including a dead zone the size of the state of New Jersey where the Mississippi River drains into the Gulf of Mexico (Picone & Van Tassel 2002). Nutrient management approaches such as cover cropping, intercropping, and diverse cropping reduce the need for added nutrients to the land (Power 2010).

Pesticide use in food production is widespread and increased dramatically in the last fifty years (WWFb). Despite efforts to decrease its use with genetic engineering, pesticide use continues to increase (Malakof and Stokstad 2013, Benbrook 2012). Pesticides, including herbicides, insecticides, and fungicides are designed to cause harm to organisms so high usage can cause unwanted harm to humans and wildlife. Short-term effects from pesticide exposure include immune reactions, organ damage, neurological damage, and death. Long-term effects include elevated cancer risks and disruption of the body's reproductive, immune, endocrine, and nervous systems (Horrigan et al. 2002). The two most commonly used herbicides for corn and soybean crops in the United States are endocrine disruptors (Horrigan et al. 2002). These chemicals cause developmental abnormalities such as impaired growth of immune systems and sexual organs. Endocrine-disrupting insecticides have been linked to extra limbs emerging from the stomach and neck of frogs (Picone & Van Tassel 2002). Such deformities are appearing in many threatened species including alligators, panthers, polar bears, and dolphins (Picone & Van Tassel 2002). Agricultural workers are particularly at risk of exposure to harmful levels and types of pesticides. Human exposure to pesticides can also come through residues in food and through contaminated drinking water and air (Horrigan et al. 2002). People consume about ten different pesticides a day when they eat foods in the United States that are commonly contaminated with pesticides (EWG 2010). Contamination of the environment is so prevalent that almost every marine organism, from the tiniest plankton to whales and polar bears, is contaminated with pesticides and other chemicals used in common consumer products (WWF). Every river sampled by the United States Geological Survey was found to contain pesticides (Gilliom et al. 2007).

There are several effective approaches to reducing the use of highly hazardous pesticides and total pesticide use. Organic production and integrated pest management (IPM) are leading alternatives. Organic production avoids the use of synthetic crop protection chemicals and instead relies on naturally-derived solutions such as mulching and crop rotations, as well as naturally-derived chemicals. It includes the management approaches that comprise IPM. IPM focuses on growing and supporting healthy crops, preventing pest build-up with natural means, and using control measures in a targeted and limited way (The World Bank). IPM uses naturally-derived as well as synthetic chemicals when needed, however, there is an aim to reduce the reliance on such chemicals. These approaches to pest management are effective and the EPA highlighted that IPM can save the producer money, improve the environment, and protect health (EPA).

## 2.4 Approaches to more sustainable agriculture

Changes at the farm-level are needed to move toward sustainable production. The best practices described earlier in the chapter (see Table 2.1) can make a difference and more are in development that can readily be adopted. The pressure to make these changes is not all on the farmer. While they are the ones implementing specific practices, the downstream players need to

**Table 2.1** Summary of sustainable agriculture practices

- Integrated pest management
- Diverse crops and varieties with high yields
- Crop rotations
- Cover crops/green cover and no/low tillage
- Managed irrigation
- Soil management/nutrient management
- Integrate conserved land or agroforestry
- No artificially heated greenhouses

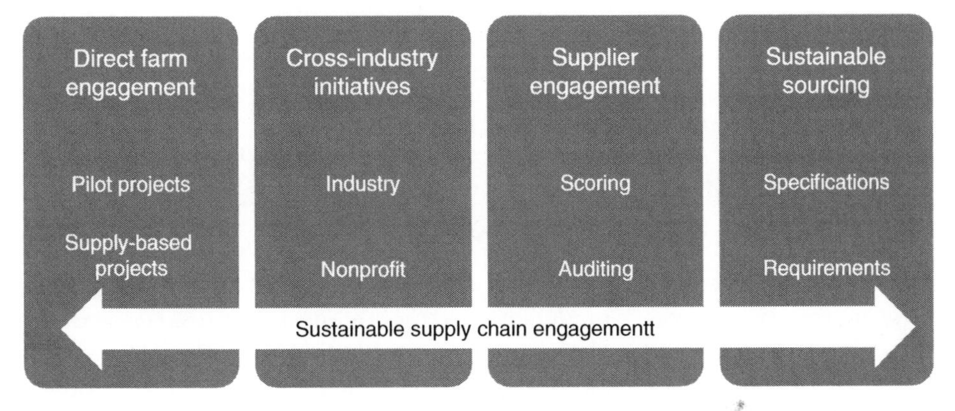


**Figure 2.1** Sustainable supply chain engagement project types (reproduced by permission of Baldwin and Atwood 2014).

encourage these changes and support their farmers. This is commonly done through sustainable supply chain efforts that engage downstream players (e.g., manufacturers, retailers, food services) in farm production.

Sustainable supply chain engagement involves a range of activities including projects directly with farms, collaborative efforts in cross-industry programs, supplier engagement, and different types of purchasing requirements (tracking to mandating) (see Figure 2.1). The best approach to begin a supply chain program depends on what would work best for the business. Typically more than one type of effort is taken on since they complement each other to lead to greater progress (see the company examples that follow). Many efforts begin with sustainable sourcing of select ingredients/materials through the purchasing process and develop from there. The cross-industry program, the Sustainable Agriculture Initiative (SAI) Platform, developed guidance on how to create and implement a sustainable supply chain program (SAI 2013). This guidance focuses more on how to use purchasing requirements/sustainable sourcing as the leading tool to engage in upstream farm activities. Sustainable sourcing can be done across the range of supplier relationships such as direct sourcing from the producer and sourcing from "anonymous" commodity markets that may require a formal standard and verification process, such as an externally developed an independently certified standard (SAI 2013).

SAI outlined the following components for a sustainable sourcing program (SAI 2013).

- Identify priority raw materials and issues to begin the program focus; based on high impacts, ability to improve, and importance to the business.
- Develop sustainability goals and requirements for the priority materials with internal or external standards.
- Implement program through the supply chain with appropriate support (e.g., capacity building) and verification (e.g., audits).
- Monitor progress by tracking, measuring, and continuously improving to meet goals.

The aim of the program should be to eventually source all of the agricultural materials from sustainable sources. Incremental progress with priority ingredients helps lead to this ultimate goal (see the Unilever example later in the chapter). This type of sourcing effort can be enhanced with other approaches including direct farm engagement, working with cross-industry initiatives, or supplier engagement efforts. One of the leading methods of supplier engagement is to survey and score suppliers on their environmental practices and performance. This works well because suppliers are surveyed for other reasons, such as legal and safety compliance, and this can fit into existing efforts and procurement processes. This and other sustainable supply chain programs deliver value through reducing costs and risks and creating opportunities for growth and program enhancement, while reducing overall environmental damage (BSR 2010). Regardless of the approach, the process for sustainable supply chain engagement is similar (see Figure 2.2).

### 2.4.1 Sustainable agriculture requirements and standards

The range of approaches from direct engagement projects on the farm to establishing purchasing requirements require a definition of sustainable agriculture or sourcing. This is typically in the form of internal or external standards. What is selected depends on the priorities and goals for the effort (see Table 2.2 for ways to evaluate the options). There may be existing standards that can be used from external organizations. The external standards should be evaluated

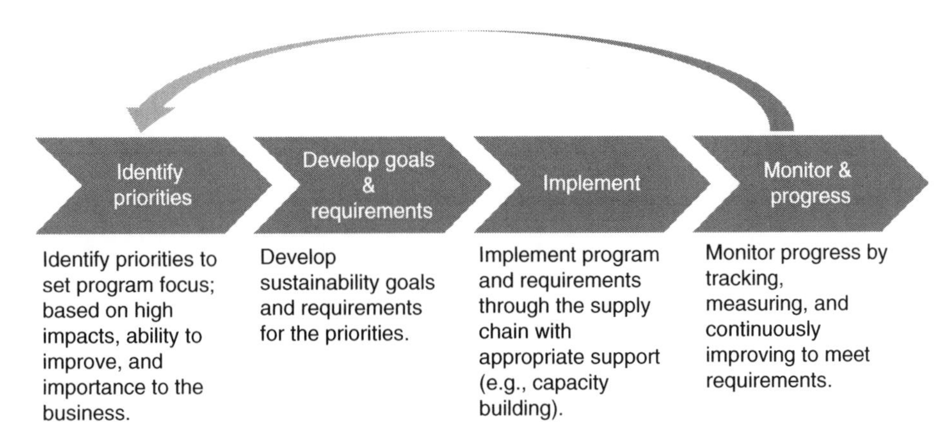


**Figure 2.2** Sustainable supply chain engagement process (reproduced by permission of Baldwin and Atwood 2014).

**Table 2.2** Sustainable sourcing standards options (reproduced by permission of Baldwin and Atwood 2014)

| | External Standards | Internal Standards |
|---|---|---|
| Selection criteria | Alignment to program goals (issue, food) and outcome (certification is not the end, it's the means) | Typically used when there isn't an external standard that meets program goals |
| Benefits | Market acceptance (alignment across companies for farmers, improved consumer awareness), verification process, rapid adoption potential (supply) | Customizable |
| Considerations | Supply costs and availability | Development and verification time and resources |

to understand if they meet the company's program goals. Many external standards are focused on one issue area or crop. If there is an external program available it can offer many benefits including established verification/certification and market acceptance (to provide alignment across companies for farmers and to improve consumer awareness). Using existing programs can also provide a fast adoption of the sourcing program if there is a supply already available that meets the requirements of the standard (see Tables 2.3 and 2.4). However, there may be supply limits or costs that reduce the feasibility of using the standard. Because of this and if there is not a standard that fits the program priorities and goals, an internal standard may need to be established. The development of an internal standard can be a long process that requires a large commitment of resources. In addition, external assistance from suppliers, farmers, and experts is needed to provide validity to the process. Using existing

**Table 2.3** Certification programs for environmentally preferable food (reproduced by permission of Green Seal and Baldwin 2012)

| Program (and the corresponding administrative organization) | Food Products Covered | Sustainability Elements Covered |
|---|---|---|
| American Humane Certified (The Humane Touch) | animal products | social |
| Animal Welfare Approved | animal products | social |
| Best Aquaculture Practices Certification (Aquaculture Certification Council) | fish, seafood | environment |
| Bird Friendly (Smithsonian National Zoological Park) | coffee | environment, social |
| Bonsucro Certification | sugar from sugarcane | environment, social |
| Certified Humane Raised and Handled (Humane Farm Animal Care) | animal products | social |
| Fair Trade (Fair Trade USA) | coffee, cocoa, food ingredients, fruits, vegetables, honey, herbs, nuts, oilseeds, spices, spirits, sugar, tea, wine | social |
| Food Alliance | temperate crops (e.g., meat, dairy, apples, etc.) | environment, social |
| Marine Stewardship Council | fish, seafood | environment |
| Protected Harvest | California strawberries, citrus, lodi winegrapes mushrooms, potatoes, stonefruit | environment |
| Rainforest Alliance | tropical crops (e.g., cocoa, coffee, fruits, tea), cattle products | environment, social |
| Roundtable for Sustainable Palm Oil | palm oil | environment, social |
| Organic (USDA National Organic Program) | most foods and beverages | environment |

frameworks can serve as a foundation to jump-start the process, such as FAO's Sustainability Assessment of Food and Agriculture Systems Guidelines (FAO 2012). Industry guidelines can also be used as a foundation. Industry efforts focused on environmental issues or sustainability are usually more applicable than generic industry programs. SAI developed a series of Principles and Practices for several crops that can be used, for example. Due to the limited options for external programs that fit well into sustainable sourcing programs,

**Table 2.4** Organic claims and certification in the united states (reproduced by permission of Green Seal and Baldwin 2012)

| 100% Organic | Organic | Made with Organic Ingredients (listing of up to three ingredients on the front of the package). | Ingredient Line Listing of Organic Ingredients |
|---|---|---|---|
|   | USDA ORGANIC | *(Cannot use UDSA Organic seal)* | *(Cannot use the word* organic *on the front panel and cannot use the UDSA Organic seal)* |
| Only contains organically produced ingredients and processing aids* | Contains at least 95% organically produced ingredients** | Contains at least 70% organic ingredients and the rest of the product cannot be GMO, irradiated, or grown with sewage sludge** | Contains less than 70% organic ingredients |

*Based on weight and excluding water and salt.

**Based on weight and excluding water and salt. May contain ingredients that are not available as USDA-certified organic as approved by the USDA National Organic Program.

Sources: USDA National Organic Program (www.ams.usda.org/).

companies often combine the use of internal standards and external standards (see the Unilever and Starbucks examples that follow).

The use of internal, external, or a combination of standards requires verification that the supply meets the requirements. This is best done through a combination of self-assessment tools for the farmer/supplier and site audits (SAI 2013). External programs often include the site audit verification through third-party certification. Third-party certification adds a cost into the supply but comes with the value of credible assurance that the supply meets the requirements. As a result, this may be most useful for issues that are challenging to verify, commodity products, or for complicated supply chains (SAI 2013). Third-party certification is useful for product labeling when such ingredients are in the product in significant amounts, since it provides independent substantiation and credibility.

## 2.4.2 Unilever sustainable agriculture program

Unilever is a global packaged food and beverage company with several brands such as Lipton Tea, Ben & Jerry's, Knorr, and Bertolli. The company has been at the forefront of advancing sustainability and as a result has a comprehensive sustainability program that includes an ambitious goal of sourcing 100% of agricultural raw materials sustainably by 2020 (Unilever a).

The company developed this goal with a ten-year outlook by setting progress milestones in phases: aiming first to source 10% by 2010 (which was met), 30% by 2012 (also met), then 50% by 2015, and finally expand to 100%

by 2020. At the end of 2012, 36% of the company's agricultural raw materials were sustainability sourced (Unilever a). The success at the early milestones was achieved, in part, by focusing on ten materials that represent two-thirds of the agricultural raw material purchases: palm oil, paper and board, soy, sugar, tea, fruit and vegetables, sunflower oil, rapeseed oil, dairy ingredients, and cocoa. However, to continue to meet the goals, Unilever expanded its focus to 30 more materials that account for about 20% of the purchases, including such things as vanilla and beef (Unilever a).

Sustainable sourcing was defined through Unilever's Sustainable Agriculture Code. The Code was developed with consultation from farmers, nonprofit organizations, and academics and includes eleven key indicators including soil management, water, and biodiversity, among others (Unilever b). Sustainable sourcing requires that production meets the Code or equivalent external certifications such as Rainforest Alliance (a Sustainable Agriculture Network standard), Fairtrade, Round Table for Sustainable Palm Oil (RSPO), Marine Stewardship Council (MSC), Forestry Stewardship Council (FSC), any organic standard recognized by the International Federation of Organic Agriculture Movements (IFOAM), and Programme for the Endorsement of Forest Certification (PEFC) for paper and board (Unilever c). Unilever relies on verification to its Sustainable Agriculture Code where there is not an existing external certification program.

Unilever purchases 12% of the global black tea and 6% of tomatoes (Unilever b). With this level of purchasing power, the importance of the sustainable sourcing program is elevated. To help advance progress, Unilever has projects targeted at these and other crops. Unilever set a goal to have all of its tea purchases be from sustainable sources by 2015. The company is on track to meet this goal by having worked directly with some of its producers in the early stages of the effort to ensure famers had the right training for sustainable cultivation. Unilever then established partnerships with Rainforest Alliance and the Sustainable Trade Initiative to expand this approach globally. In addition to the tea being cultivated sustainably, producers have experienced improved yields through this effort (Unilever d). Unilever also works closely with farmers in the Knorr Sustainability Partnership to develop approaches and accelerate sustainability advances. In Greece, Knorr provided tensiometers to sun-ripened tomato growers so they can measure soil moisture in order to avoid under- and overwatering (Unilever e). This approach has led to improved yields and water conservation to advance sustainable production (Unilever e).

Unilever strives to continuously improve its approach in sustainable sourcing through collaborative efforts. Unilever has a Sustainable Sourcing Advisory Board of external experts from academia and nonprofit organizations to help advance the company's program (Unilever f). Unilever is also active in cross-industry initiatives that support sustainable sourcing such as the Sustainable Agriculture Initiative Platform. These combined approaches have enabled Unilever's success in sustainability and continued leadership.

### 2.4.3 Starbucks C.A.F.E practices

Starbucks is a global coffee retailer with 18,000 stores across 60 countries. Starbucks developed a comprehensive sustainability program that includes community, operational, and supply chain strategies. The supply chain efforts are focused on ethically purchased and responsibly produced coffee, tea, cocoa, and manufactured goods (Starbucks).

Starbucks set a goal for 100% of coffee to be ethically sourced by 2015. By 2012, 509 million pounds of coffee or 93% was ethically sourced, including 90% that met their internally-developed standard (the rest met Fair Trade standards) (Starbucks 2012). Their Coffee and Farmer Equity (CAFE) Practices standard addresses social, economic, and environmental criteria. There are more than 200 indicators in the standard with soil management, water quality protection and conservation, energy conservation, biodiversity preservation, and agrochemical reduction comprising some of elements of the environmental criteria. Soil management indicators, for example, include requirements for vegetative cover for slopes, biological fertilization with leguminous crops, soil analysis for nutrients and organic matter, and shade canopy (Starbucks 2013). Starbucks has similar internal standards for cocoa and works with the Ethical Tea Partnership for tea. Each program includes independent verification that the supply meets the company's purchasing requirements.

Starbucks also has a partnership with Conservation International to address risks to farming communities in their supply chain. The partnership has helped identify strategies for improving the sustainability of coffee production processes, the conservation and restoration of natural habitat, and opportunities to facilitate farmer access to forest carbon markets and other forms of assistance. This effort provided reforestation and restoration that led to avoided emissions of 50,000 tons of carbon.

### 2.4.4 Walmart sustainability index

Walmart, the world's largest retailer, has been advancing sustainability since 2006. The company has goals for achieving zero waste and using 100% renewable energy. There is also focused effort to "sell products that sustain people and the environment (Walmart)."

A key tool that the company uses to progress on this goal is the Walmart Sustainability Index. The Index is a set of category-specific questions about sustainability performance on top issues. The questions were developed through the science-based, multi-stakeholder effort at The Sustainability Consortium (TSC). Walmart's suppliers answer the questions every year. The Index was launched in 2012 and reached 190 of Walmart's product categories within 1 year (Baldwin 2013).

Supplier responses are scored and the information is provided to Walmart's merchant team, the staff that purchases the products for the stores. The scorecards are used to uncover potential areas for focused projects to drive improvement. These can be through a partnership between Walmart and its suppliers or could be a more broad industry effort. Walmart has developed a number of projects aimed to improve the sustainability of its products. For example, when pasta, cereal, and other grain-based suppliers were scored, fertilizer use was highlighted as a key improvement opportunity. Walmart then engaged its suppliers and industry groups to begin to advance adoption of improved practices through connecting farmers with tools and approaches to optimize fertilizer use. The effort is expected to reduce fertilizer needs by 30% leading to a GHG emission reduction of 7 million metric tons of carbon dioxide equivalents by 2020 (Gallagher 2013).

### 2.4.5 PepsiCo sustainable farming initiative

PepsiCo is a leading food and beverage company, well known for its carbonated beverages under the Pepsi brand and other brands such as Frito-Lay, Tropicana, Quaker, and Gatorade. In 2012, the company launched the Sustainable Farming Initiative (SFI) (PepsiCo a). The program is focused on potato, citrus, oats, rice, and corn crops with the aim to decrease risk for farmers and to build resilience in the supply chain by improving agricultural practices, especially to reduce on-farm water use and GHG emissions (Albanese 2012).

SFI is an internally-developed program that measures nine environmental, four social, and three economic indicators. Agrochemicals, air, biodiversity, energy, GHGs, nutrients, soil, waste, and water are the environmental indicators for agricultural operations. By measuring these indicators and comparing to a baseline, more visibility is available for project development and continuous improvement (PepsiCo 2012). The criteria are aligned with existing programs including Global G.A.P. and Rainforest Alliance (PepsiCo 2012).

By the end of 2012, 36 pilot projects were completed to kick-off the program (PepsiCo 2013). One set of pilots included a web-based tool, "iCrop," that helps farmers optimize productivity and water use and reduce carbon emissions. These pilots helped U.K. potato farmers increase yields by 13% and reduce water use by 8%. Another project that began prior to the launch of SFI helped develop a direct seeding approach for rice that avoids the typical water-intensive practices. This approach saved 7 billion liters of water and had a 70% reduction in GHGs (PepsiCo b).

### 2.4.6 Sysco Corporation's sustainable agriculture/IPM initiative

Sysco Corporation is one of the largest food service distributors in the United States. Sysco began its sustainable agriculture program in 2004 with a focus on fruit and vegetable production. The aim is to maintain a safe food supply while working toward the prudent use of inputs including fertilizers, pesticides, energy, and water (Sysco a). This is being done through the development of a leading integrated pest management (IPM) program that also addresses fertilizer, water, and waste. Sysco developed its sustainable agriculture/IPM standard and applies it to all of its canned and frozen fruit and vegetable products that are grown across the globe. Assistance in developing the standard was provided by the IPM Institute and was modeled off of the Wegman's grocery store program in New York State (Sustainable Food Lab 2010).

Each year a third-party, on-site audit is conducted of supplier processing facilities and raw material suppliers. For example, suppliers are scored on their progress on implementing IPM including understanding key pests, identifying and implementing multiple strategies to prevent pest damage, systematic pest and crop monitoring, use of thresholds to determine when control actions are economically justified, managing pest resistance to pesticides, managing drift and environmental emergencies, selecting the least-toxic pest control options, and tracking and setting goals for pesticide use and hazard reduction (Sysco b). In addition, suppliers annually report environmental indicator information such as pesticide and nutrient applications. The program has proven to be successful. In 2011, suppliers avoided the use of more than 341,600 pounds of pesticides, 6 million pounds of fertilizer, 714 million gallons of water, and 525,000 gallons of fuel (Sysco a).

## 2.5 Summary

Agricultural production is a leading contributor to the environmental damage caused by the food system. Climate impacts, land and biodiversity damage, and water and pollution issues have been growing challenges. Existing best practices offer the opportunity to reduce the overall burden and have demonstrated to improve yields and decrease costs (see Table 2.1). Additional advancements can shift the relationship food production has with the environment to be beneficial and more sustainable. Each player in the food supply can facilitate this progress, even if a company is not directly engaged with agricultural production. Sustainable sourcing efforts are a growing practice for food manufacturers, retailers, and food services. With effective sustainable sourcing programs, the entire supply chain can collaborate to make the needed changes in agricultural production to avoid the environmental damage with the current system and work to rebuild the land and protect biodiversity and water (see Table 2.5).

## Resources

SAI Platform: www.saiplatform.org
FAO: www.fao.org

**Table 2.5** Summary of the principles–practices–potential of sustainability for agricultural production

| | |
|---|---|
| **Principle** | *Agricultural production beneficially contributes to the environment while efficiently using natural resources and maintaining healthy climate, land, water, and biodiversity.* |
| **Practices** | • Implement responsible production techniques such as:<br>  • Integrated pest management<br>  • Diverse crops and varieties with high yields<br>  • Crop rotations<br>  • Cover crops/green cover and no/low tillage<br>  • Managed irrigation<br>  • Soil management/nutrient management<br>  • Integrate conserved land or agroforestry<br>  • No artificially heated greenhouses<br>• Source and support responsibly produced crops |
| **Potential** | • Replenish water<br>• Revitalize biodiversity<br>• Build soil fertility<br>• Restore land and provide a carbon sink |

# References


Albanese, M. (2012). How she leads: Beth Sauerhaft of PepsiCo. Retrieved from http://www.greenbiz.com/blog/2012/10/16/how-she-leads-beth-sauerhaft-pepsico?page=0%2C0

Baldwin, C. (2012). *Greening food and beverage services: A green seal guide to transforming the industry.* Lansing, MI: American Hotel and Lodging Educational Institute.

Baldwin, C. (2013). How Walmart is transforming dairy sustainability. *Environmental Leader.* Retrieved from http://www.environmentalleader.com/2013/08/14/how-walmart-is-transforming-dairy-sustainability/

Baldwin, C., & Atwood, R. (2014). *Pure strategies webinar: Supply chain engagement for sustainable change* Sustainability Management Association, January 23, 2014.

Bellarby, J., Foereid, B., Hastings, A., & Smith, P. (2008). *Cool farming: Climate impacts of agriculture and mitigation potential.* Retrieved from http://www.greenpeace.org/international/Global/international/planet-2/report/2008/1/cool-farming-full-report.pdf

Benbrook, C. (2012). Impacts of genetically engineered crops on pesticide us in the U.S.—the first sixteen years. *Environmental Sciences Europe,* 24. doi:10.1186/2190-4715-24-24. Retrieved from http://www.enveurope.com/content/24/1/24

BSR. (2010). *The business case for supply chain sustainability: A brief for business leaders.* Retrieved from http://www.bsr.org/reports/Beyond_Monitoring_Business_Case_Brief_Final.pdf

Convention on Biological Diversity (CBD). (2010). *Global biodiversity outlook 3.* Retrieved from http://www.cbd.int/doc/publications/gbo/gbo3-final-en.pdf

Cox, T. S., & Jackson, W. (2002). Agriculture and biodiversity: Genetic engineering and the second agricultural revolution. In N. Eldredge (Ed.), *Life on earth: An encyclopedia of biodiversity, ecology, and evolution* (pp. 96–99). Santa Barbara, CA: ABC-CLIO/Greenwood.

Drinkwater, L. E., & Snapp, S. S. (2007). Nutrients in agroecosystems: Rethinking the management paradigm. *Advances in Agronomy,* 92, 163–186. doi:10.1016/ S0065-2113(04)92003-2

Environmental Working Group (EWG). *EWG's shoppers guide to pesticides in produce.* Retrieved from http://www.ewg.org/foodnews/

Food and Agriculture Organization of the United Nations (FAO). (1995). *Water: A finite resource in dimensions of need: An atlas of food and agriculture.* Retrieved from http://www.fao.org/docrep/u8480e/U8480E0c.htm#Irrigation

Food and Agriculture Organization of the United Nations (FAO). (2006). *Livestock's long shadow: Environmental issues and options.* Retrieved from http://www.fao.org/docrep/010/a0701e/a0701e00.HTM

Food and Agriculture Organization of the United Nations (FAO). (2011). *The state of the world's land and water resources for food and agriculture: Managing systems at risk.* Retrieved from www.fao.org/nr/solaw/solaw-home/en

Food and Agriculture Organization of the United Nations (FAO). (2012). *Sustainability assessment of food and agriculture systems (SAFA) guidelines: Draft 4.0 compact version.* Retrieved from http://www.fao.org/fileadmin/user_upload/suistainability/SAFA/SAFA_Guidelines_draft_Jan_2012.pdf

Food and Agriculture Organization of the United Nations (FAO). (2013). *Sustainable sourcing of agricultural raw materials: A practitioner's guide.* Retrieved from http://www.saiplatform.org/uploads/documents/Sustainable_Sourcing_Guide-SAI_Platform.pdf

Gallagher, T. (2013). *A retail revolution in responsibility*, Retrieved from http://blog.purestrategies.com/blog/bid/317829/A-Retail-Revolution-in-Responsibility

Gallai, N., Salles, J. M., Settele, J., & Vaissiere, B. E. (2009). Economic valuation of the vulnerability of world agriculture confronted with pollinator decline. *Ecological Economics,* 68, 810–821.

Galloway, J. N., et al. (2004). Nitrogen cycles: Past, present, and future. *Biogeochemistry,* 70, 153–226. doi:10.1007/s10533-004-0370-0

Gilliom, R., et al. (2007) *Pesticides in the nation's streams and ground water, 1992–2001.* Retrieved from http://pubs.usgs.gov/circ/2005/1291/

Horrigan, L., Lawrence, R., & Walker, P. (2002). How sustainable agriculture can address the environmental and human health harms of industrial agriculture. *Environmental Health Perspectives*, 110(5), 445–456.

Institute of Medicine. (2014). *Sustainable diets: Food for healthy people and a healthy planet—workshop summary.* Retrieved from http://iom.edu/Reports/2014/Sustainable-Diets-Food-for-Healthy-People-and-a-Healthy-Planet.aspx?utm_medium=etmail&utm_source=National%20Academies&utm_campaign=February+2014&utm_content=Web&utm_term

Intergovernmental Panel on Climate Change (IPCC). (2007). *IPCC fourth assessment report: Climate change 2007. Section TS2.5.* Retrieved from http://www.ipcc.ch/publications_and_data/ar4/wg1/en/tssts-2-5.html

International Food Policy Research Institute. (2009). *Climate change: Impact on agriculture and costs of adaptation.* Retrieved from http://www.ifpri.org/sites/default/files/publications/pr21.pdf

Lappé, A. (2010). *Diet for a hot planet: The climate crisis at the end of your fork and what you can do about it.* New York: Bloomsbury.

Malakof, D., & Stokstad, E. (2013) Pesticide planet. *Science*, 341 (6147), 730–731. DOI: 10.1126/science.341.6147.730

National Geographic. (2010). *American lifestyle costs nearly 2000 gallons of water each day.* Retrieved from http://blogs.nationalgeographic.com/blogs/news/chiefeditor/2010/08/american-lifestyle-costs-nearl.html

National Geographic. (2011). *The hidden water we use.* Retrieved from http://environment.nationalgeographic.com/environment/freshwater/embedded-water/

Nierenberg, D., & Reynolds, L. (2012). *Innovation in sustainable agriculture: Supporting climate-friendly food production.* Retrieved from http://www.worldwatch.org/bookstore/publication/worldwatch-report-188-innovations-sustainable-agriculture-supporting-climate-f

Oerke, E-C. (2006). Crop losses to pests. *Journal of Agricultural Science,* 144, 31–43. Retrieved from http://www.nrel.colostate.edu/ftp/conant/SLM-proprietary/Oerke_2006.pdf

PepsiCo a. *Sustainable agriculture.* Retrieved from http://www.pepsico.com/Purpose/Environmental-Sustainability/Agriculture

PepsiCo b. *Our commitment to sustainable agriculture practices: Good for business, good for society.* Retrieved from http://www.pepsico.com/Download/PepsiCo_agri_0531_final.pdf.

PepsiCo. (2012). *Presentation from the 2012 World Water Week in Stockholm.* Retrieved from http://www.worldwaterweek.org/documents/WWW_PDF/2012/Thur/The-importance-of-farm/Ian_HJ.pdf

PepsiCo. (2013). *PepsiCo recognized for its sustainability efforts.* Retrieved from http://www.marketwatch.com/story/pepsico-recognized-for-its-sustainability-efforts-2013-09-12

Picone, C., & Van Tassel, D. (2002). Agriculture and biodiversity loss: Industrial agriculture. In N. Eldredge (Ed.), *Life on earth: An encyclopedia of biodiversity, ecology, and evolution* (pp. 99–105). Santa Barbara, CA: ABC-CLIO/Greenwood.

Pimentel, D., Williamson, S., Alexander, C. E., Gonzalez-Pagan, O., Kontak, C., & Mulkey, S. E. (2008). Reducing energy inputs in the US food system. *Human Ecology,* 36, 459–471. doi: 10.1007/s10745-008-9184-3

Power, A. (2010). Ecosystem services and agriculture: Trade-offs and synergies. *Philosophical Transactions of The Royal Society,* 365, 2959–2971.

Reynolds, L., & Nierenberg, D. (2012) *Innovations in sustainable agriculture: Supporting climate-friendly food production.* Retrieved from http://www.worldwatch.org/system/files/188%20climate%20and%20ag_FINAL.pdf

Rost, S., Gerten, D., Hoff, H., Lucht, W., Falkenmark, M., & Rockstrom, J. (2009) Global potential to increase crop production through water management in rain-fed agriculture. *Environmental Research Letters,* 4. doi:10.1088/1748-9326/4/4/044002. Retrieved from http://iopscience.iop.org/1748-9326/4/4/044002/fulltext/

Shand, H. (2000). *Biological meltdown: The loss of agricultural biodiversity.* Retrieved from http://www.urbanhabitat.org/node/921

Starbucks. *Ethical sourcing.* Retrieved from http://www.starbucks.com/responsibility/sourcing

Starbucks. (2012). 2012 *Global responsibility report: Year in review.* Retrieved from http://globalassets.starbucks.com/assets/581d72979ef0486682a5190eca573fef.pdf

Starbucks. (2013). *C.A.F.E. practices: Generic scorecard.* Retrieved from http://www.scsglobalservices.com/files/CAFE_SCR_Genericv3.1_012513_0.pdf

Sustainable Agriculture Initiative (SAI) Platform. (2013). *Sustainable sourcing of agricultural raw materials: A practitioner's guide.* Retrieved from http://www.saiplatform.org/sustainable-sourcing-guide

Sustainable Food Lab. (2010). IPM @ Sysco. Retrieved from http://www.sustainablefood.org/business-coalition/59-imp-at-sysco

Sysco a. *Sustainable agriculture.* Retrieved from http://sustainability.sysco.com/supplying-food-responsibly/sourcing-food-responsibly/sustainable-agriculture/

Sysco b. *SYSCO Corporation's PESP strategy.* Retrieved from http://www.epa.gov/pesp/pesp/members/strategies/sysco_corp.pdf

Unilever a. *Our strategy and footprint.* Retrieved from http://www.unilever.com/sustainable-living/sustainablesourcing/why/index.aspx

Unilever b. *Responsible and sustainable sourcing: Standards guide for our supply partners*. Retrieved from http://www.unilever.com/images/Unilever%20Responsible%20and%20Sustainable%20sourcing_tcm13-265398.pdf

Unilever c. *Certification vs. self-verification*. http://www.unilever.com/aboutus/supplier/sustainablesourcing/sustainableagriculturecode/index.aspx

Unilever d. *Sustainable tea*. Retrieved from http://www.unilever.com/sustainable-living/sustainablesourcing/tea/

Unilever e. *Gastouni*. Retrieved from http://www.unilever.com/images/Gastouni_tcm13-286142.pdf

Unilever f. *Agricultural sourcing partnerships*. Retrieved from http://www.unilever.com/sustainable-living/sustainablesourcing/agriculturalsourcingpartnerships/index.aspx

U.S. Department of Agriculture (USDA). 2012. Report on the national stakeholders conference on honey bee health: National honey bee health stakeholder conference steering committee. Retrieved from http://www.usda.gov/documents/ReportHoneyBeeHealth.pdf

U.S. Environmental Protection Agency (EPA). *Integrated pest management*. http://www.epa.gov/agriculture/tipm.html

U.S. Environmental Protection Agency (EPA). (1998). *Reducing water pollution from animal feeding operations: Testimony before subcommittee on forestry, resource conservation, and research of the committee on agriculture, U.S. House of Representatives, 13 May 1998*. Retrieved from http://www.epa.gov/ocirpage/hearings/testimony/105_1997_1998/051398.htm

Walmart. *Environmental sustainability*. Retrieved from http://corporate.walmart.com/global-responsibility/environment-sustainability

Wines, M. (2013). Wells dry, fertile plains turn to dust. *The New York Times*. Retrieved from http://www.nytimes.com/2013/05/20/us/high-plains-aquifer-dwindles-hurting-farmers.html?pagewanted=all&_r=0

The World Bank. *Pest management guidebook*. Retrieved from http://web.worldbank.org/WBSITE/EXTERNAL/TOPICS/EXTARD/EXTPESTMGMT/0,,contentMDK:20631451 ~ menuPK:1605318 ~ pagePK:64168445 ~ piPK:64168309 ~ theSitePK:584320,00.html

World Economic Forum (WEF). (2012). What if the world's soil runs out. *Time*. Retrieved from http://world.time.com/2012/12/14/what-if-the-worlds-soil-runs-out/

World Health Organization (WHO). *Biodiversity*. Retrieved from http://www.who.int/globalchange/ecosystems/biodiversity/en/

World Wildlife Fund (WWF a). *Facts about our earth*. http://wwf.panda.org/about_our_earth/

World Wildlife Fund (WWF b). Pollution. Retrieved from http://worldwildlife.org/threats/pollution

World Wildlife Fund (WWF). (2012). *The 2050 criteria: Guide to responsible investment in agricultural, forest, and seafood commodities*. http://awsassets.panda.org/downloads/the_2050_critera_report.pdf

# 3 Welfare and Environmental Considerations in Production and Harvesting of Animals, Fish, and Seafood

*Principle: Use of animals, fish, and seafood in the food supply optimizes their well-being and adds to environmental health.*

Our food supply relies on ingredients and foods from animals, fish, and seafood—from fresh meat to isolated dairy protein. With the growth in population and improvement in standards of living, this demand will increase. There are unique sustainability considerations for the production and harvesting of these products that includes the well-being of the animals and their disproportionately large impact to the environment. These issues carry tremendous risk into the food supply so businesses and industry groups have increasingly become involved in addressing them and strengthening the contribution animals, fish, and seafood have in a sustainable and thriving food system.

## 3.1 Livestock care

More than 70% of food companies recognize the business importance of animal welfare (Amos & Sullivan 2012). This is because animals that are treated well, with their natural behavioral and physical needs met, are healthier and safer sources of food (Pew Charitable Trusts 2008). However, there is greater demand to produce food, so some of the approaches used to find efficiencies in

*The 10 Principles of Food Industry Sustainability*, First Edition. Cheryl J. Baldwin.

farm animal production have had detrimental effects on the welfare of the animals (Reed 2011). Consumers are growing concerned about the treatment of animals in the food supply (American Humane Association 2013). Welfare concerns include animal abuse and neglect, restriction or boredom, and problematic management approaches that create new issues or compound others (e.g., sub-therapeutic antibiotic use with farm animals contributing to antibiotic resistance in humans). Some of the issues come from poor handling, but many arise from animals raised in systems that are designed for efficiency. Close confinement of animals, for example, may require them to be physically altered to avoid injury (e.g., hen beak clipping, hog tail docking). The housing size limits can be so extreme that the animals cannot perform natural behaviors (turn around, stretch wings). Food industry action to address the issues is growing but is in its early stages, with less than 50% of companies with an animal welfare policy and just 25% with objectives and goals (Amos & Sullivan 2012).

Part of the challenge with addressing animal welfare issues is that they have both ethical (value-based) and scientific (empirical) components (Pew Charitable Trusts 2008). There is general agreement that responsible animal care addresses at least what is referred to as the Five Freedoms (see Figure 3.1):

- *Freedom from hunger or thirst* by ready access to freshwater and a diet to maintain full health and vigor.
- *Freedom from discomfort* by providing an appropriate environment including shelter and a comfortable resting area.

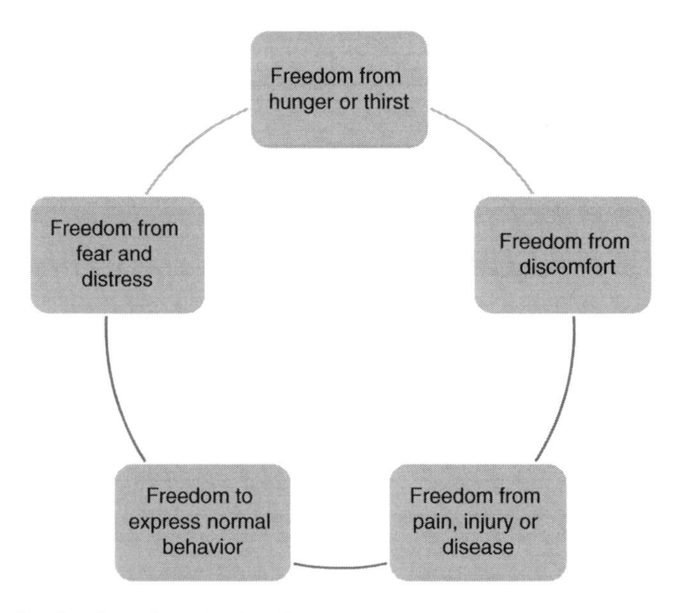


**Figure 3.1** Five freedoms for animal welfare.

- *Freedom from pain, injury, or disease* by prevention or rapid diagnosis and treatment.
- *Freedom to express normal behavior* by providing sufficient space, proper facilities, and company of the animal's own kind.
- *Freedom from fear and distress* by ensuring conditions and treatment that avoid mental suffering.

In addition, engineered solutions to address management considerations (e.g., efficiency) should not create new problems.

The reality of today's animal production systems is that the conditions can be confining, overcrowded, barren, and at times unsanitary environments. Companies are starting to address some of these top animal welfare issues including cages for egg-laying hens and gestation crates for sows (Humane Society of the U.S. 2013). Hen cages and sow crates are commonly used housing systems in the United States, but these enclosures can be so small that the bird or animal is not able to express normal behaviors such as walking or turning around, one of the five freedoms. Temple Grandin, a leading animal welfare expert from Colorado State University, calls for housing that allows for normal behaviors and movement where the bird or animal can lie down, turn around, and easily move (Grandin). It is inhumane to prevent the animal from having normal behaviors and movement (Pew Charitable Trusts 2008). A number of states in the United States have begun to ban some of these housing systems, along with the United Kingdom and European Union.

Emerging animal well-being issues include engineered solutions used on the farm, such as growth-promoting agents including beta agonists, hormones, and antibiotics. Antibiotics are sometimes used to reduce infections typically caused from underlying care issues of poor sanitation or overcrowding where germs spread easily and to improve the animal's productivity and profitability (e.g., sub-therapeutic antibiotic use in feed). While antibiotics are needed for therapeutic uses, these other conditions have contributed to agriculture accounting for 70 to 80% of antibiotics sold in the United States—with the pork industry emerging as one of the heaviest users, especially of the medically important antibiotics used to treat humans (Forini, et al. 2005). Data from the U.S. Food and Drug Administration (FDA) show that the use of antibiotics for meat and poultry production has reached record levels and is increasing (FDA 2013; Pew Charitable Trusts 2013).

Development of antibiotic-resistant microorganisms and less effective medicines have been the leading concerns with the high levels of antibiotic use in agriculture. These issues are globally prevalent and growing, especially in areas of high antibiotic use such China's pork production that uses very high level of antibiotics and appears to be rapidly creating antibiotic resistant bacteria (Charles 2013). Several classes of antibiotics are used both in livestock and humans, such as penicillins and tetracyclines. As a result, the

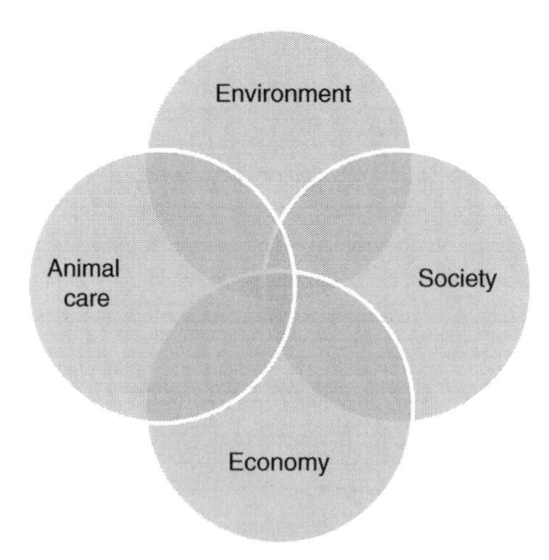


**Figure 3.2** Sustainability pillars for animal-based food production (Baldwin 2013).

World Economic Forum, the World Health Organization, and the U.S. Centers for Disease Control and Prevention have called for reduced use of such antibiotics in agriculture to help minimize the potential that the medicines needed for people become ineffective (Centers for Disease Control and Prevention 2013; World Economic Forum 2013; World Health Organization 2012). Good management with proper sanitation and husbandry practices that keep the animals well-nourished and healthy (e.g., environment, weaning practices), in combination with addressing the five freedoms, can help address this issue.

What is at risk with animal care issues? The drivers of corporate sustainability (see the introductory chapter) outline the risks. Consumers are growing more concerned about farm animal welfare (American Humane Association 2013). Brand reputation and trust are critically vulnerable if a company is found to be affiliated with poor animal welfare practices, which may impact consumer and customer perceptions and growth opportunities. Food retailers and food service companies are developing policies on animal welfare so suppliers that do not meet these can lose business. Animal welfare is a subject of emerging regulations that can further add to cost risks. Proactive companies have realized this situation and have begun to advance their sustainability programs by including animal care as a consideration (see Figure 3.2).

## 3.1.1 Approaches to address livestock welfare

The Business Benchmark on Farm Animal Welfare (BBFAW) evaluates the food industry's progress in addressing animal welfare (Amos & Sullivan 2012). While there remains significant work ahead, an overall progression to addressing the issues has emerged, see Figure 3.3. Most commonly, efforts are minimal and aimed to meet regulatory compliance and rely on voluntary

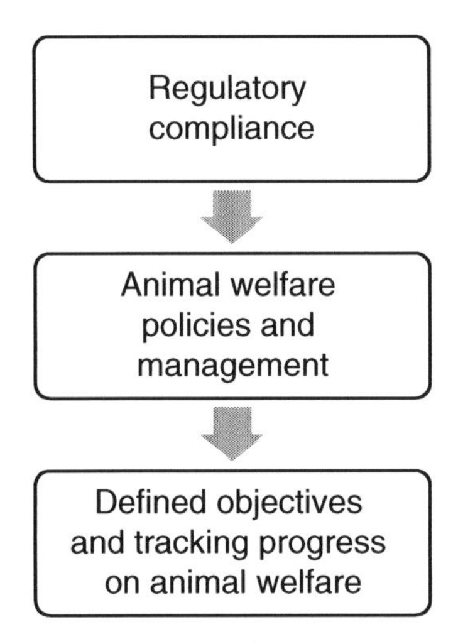


**Figure 3.3** Livestock welfare strategy progression.

industry standards of practice to address animal welfare with a very limited scope. A minority (but a growing number) of companies are emerging with policies and management systems to address the issues broadly and more strategically. There are a few leaders at the forefront that have clearly defined objectives that they track progress against. Ben & Jerry's Caring Dairy program stands out with on-farm sustainability and welfare indicators being measured and the company provides assistance to the farmer to develop an action plan to improve practices.

The BBFAW highlighted the following best practices in addressing animal welfare (Amos & Sullivan 2012):

- A clear understanding of the business case for action, underpinned by a robust assessment of the business risks and opportunities.
- An explicit farm animal welfare policy that sets out their core principles and beliefs on farm animal welfare, and that explains how these beliefs are addressed and implemented throughout the business.
- Detailed policies on specific farm animal welfare-related issues including (as relevant) policies on:
    - The close confinement of livestock, including formal statements of their positions on the use of sow stalls, farrowing crates, battery cages, tethering, and veal crates.
    - The use of genetically modified or cloned animals or their progeny.
    - Routine mutilations (e.g., teeth clipping, tail docking, dehorning, disbudding, mulesing, and beak trimming).

  - The use of antibiotics and hormone growth promoters.
  - Pre-slaughter stunning.
  - Long-distance transport of live animals.
- A clear understanding of the strengths and weaknesses in their approach to animal welfare, and an action plan for addressing weaknesses.
- Clearly defined objectives and targets for farm animal welfare.
- Established processes for monitoring implementation of the policy (both within their own operations and in their supply chains), for taking action in the event that problems arise, and for capturing and reflecting on innovations and improvements.
- Assigned responsibilities for farm animal welfare issues, at board/senior management and operations levels.
- Staff competencies to effectively manage farm animal welfare.
- Incentivize suppliers to achieve high standards of farm animal welfare performance.
- Incorporate animal welfare criteria into supply chain audits and developed improvement plans with key suppliers.
- Report publicly on their animal welfare objectives, status, and progress.
- Actively supported and participated in research and development programs to address high priority impacts.
- Promote higher farm animal welfare to customers through marketing and communication activities designed to drive up demand for higher welfare products.

This approach mirrors the sustainable supply chain engagement discussed in the previous chapter (summarized in Figure 3.4). A key best practice is the use of criteria that can be verified through the supply chain. Since there are not universally accepted standards available yet, companies have developed their own or work with suppliers to align on the criteria (see Tesco example that follows). The BBFAW recommends the use of input and outcome measures for the criteria. Inputs such as straw and perches for hens to have access to normal behaviors is useful, but not enough to understand if the welfare of the bird is better off than without it. As a result also looking at health and well-being outcomes including illness and behavior are important to evaluate. Animal welfare expert Dr. Temple Grandin has outlined the following outcome-based measures as important starting points for a

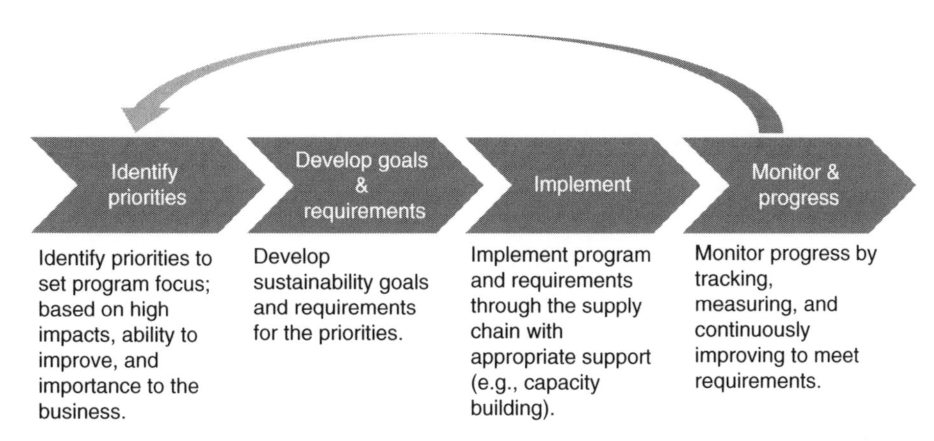


**Figure 3.4** Sustainable supply chain engagement process (reproduced by permission of Baldwin & Atwood 2014).

good program, to be expanded upon for the specific animal and production environment (Grandin):

- Body condition index
- Poor body condition (skinny)
- Lameness
- Lesions and injuries
- Coat/feather condition—coat condition is an important measure in organic systems to detect untreated parasites
- Animal cleanliness
- Culling rate and longevity
- Ammonia levels in indoor housing
- Mortality
- The animal or bird is able to lie down, turn around, and easily move in the enclosure. All animals or birds should have enough space so they can all lie down at the same time without being on top of each other.
- Symptoms of heat or cold stress

Another approach companies take is to work directly with suppliers to ensure appropriate animal care practices are implemented. Regardless of the approach, a set of criteria used across the supply or supplier partnerships and verification that the requirements are met is critical (see Figure 3.5).

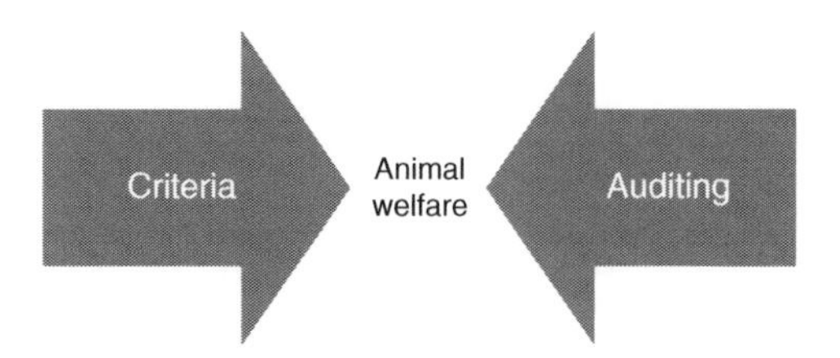


**Figure 3.5** Interaction of criteria and auditing.

Independent verification is the most robust option, but may have a high cost hurdle. There are organizations that provide this kind of service, such as Validus. There is also a limited supply of livestock products that already meet independent standard and certification programs such as Certified Humane Raised and Handled, Animal Welfare Approved, Global Animal Partnership, and Animal Humane Certified.

### 3.1.1.1 *Unilever's animal welfare approach*

Global food manufacturer, Unilever, has committed to sourcing 100% of its ingredients from sustainable sources by 2020. This was highlighted in the previous chapter about agricultural production and also applies to animal-based foods. Unilever's animal welfare program was rated highly by the BBFAW since the company aims to source products with good animal welfare practices that address issues such as housing, hygiene, feeding and feed, health management and the management of antibiotics, water supply, mutilations, transport, slaughtering practices and traceability. For Unilever this translates to monitoring and improving animal welfare based on the five freedoms, as well as legal compliance and continuous improvement (Amos & Sullivan 2012). To advance on this approach the company began its focus on eggs and dairy products. The company has already made notable progress with sourcing about one-third of North American needs from cage-free eggs and building a strong sustainable dairy program that provides about one-third of dairy ingredients (Unilever 2012). The dairy program includes the following criteria (more on the company's Caring Dairy program will be discussed later in the chapter) (Ben & Jerry's a):

- Use of herd health plan, monitoring, and protocols
- Calves
- Animal husbandry and safety training
- Cow nutrition
- Animal monitoring
- Locomotion
- Sanitation

- Body condition scores and hock lesions
- Milk quality
- Cull rate
- Animal environment
- Facilities
- Handling and transport
- Special needs animals

#### *3.1.1.2 Tesco's animal welfare program*

Tesco, a global food retailer developed a leading animal welfare program. Tesco's animal welfare program works with its suppliers to identify any issues and work toward improvements using criteria from its Code of Practice (Tesco). The Code of Practice was developed by Tesco and includes input and outcome measures (Compassion for World Farming, Tesco). Tesco's outcome measures go beyond baseline indicators, such as those noted in the list from Dr. Temple Grandin, to also include more comprehensive measures such as behavior (e.g., physical activity) and management requirements (e.g., antibiotic usage). Tesco implemented this program across the supply chain and has seen clear improvement in animal welfare. The chickens in Tesco's supply chain were found to have fewer illnesses and injuries and also were more physically active and stronger (Compassion for World Farming). The program works through the entire supply chain with suppliers submitting information to Tesco on a monthly basis. Best practices are shared across the supply chain to encourage continuous improvement. Tesco found that this thorough approach provides the desired outcomes in animal welfare through the supply chain and it also benefits their producers.

## 3.2 Fish and seafood

The world's main fishing grounds are now being fished at or above their sustainable limits (Food and Agriculture Organization of the United Nations [FAO] 2004). Over 75% of the world's fish are fully exploited or overexploited and 90% of all large fishes have disappeared from the world's oceans (Convention on Biological Diversity [CBD] 2010). There is significant concern that such rapid extinction is likely to lead to broader collapses of ecosystems at a global scale, threatening food supplies for hundreds of millions of people (CBD 2010). Poor fisheries management is the largest threat to ocean life and habitats and the livelihoods supported by fishing (WWF).

In Newfoundland, Canada, cod was a key harvest for the area (employing 110,000 people) and the stock seemed inexhaustible (WWF). However, cod was fished to commercial extinction in the 1990s and the ecosystem in the waters was altered to such an extent that cod still has not recovered, if it ever will (WWF). Despite this often cited example, regulatory controls are still vastly insufficient with the global fleet of fishing boats estimated to be enough to catch two-and-a-half times what the ocean produces. Inadequate laws and enforcement is widespread, with about 20% of the supply caught in the wild coming from illegal and unregulated sources (Pew Charitable Trusts).

The ecosystem is affected not just by the loss of the species, but many fishing practices also harm the aquatic environment. Bottom trawling for fish by dragging large nets along the sea floor is one of the most damaging fishing methods. Bottom trawling destroys the ocean habitat and also catches nontarget species, called bycatch, including sea turtles, dolphins, and sharks. Bycatch is a problem because the animals get injured or killed in the process and represent a notable portion of the catch, about 8% (Monterey Bay Aquarium 2011). Better management of fishing and using more responsible catching methods, such as pole and line fishing or hook and line fishing, can avoid these problems.

The Blue Ocean Institute and the Monterey Bay Aquarium have studied these issues and developed five criteria to evaluate and rate the sustainablity of fish. The ratings are compiled into lists of more responsible options (sometimes referred to as "green") and options that should be avoided (sometimes referred to as "red"). The criteria include the following:

1. Inherent vulnerability/life history—growth rate of the fish and how quickly the fish reproduces.
2. Abundance—the status of wild stocks compared to natural or unfished levels.
3. Habitat quality and fishing gear impact—the effect of fishing practices on habitats and ecosystems.
4. Management—the effectiveness of the regulations to protect the fish and their ecosystem.
5. Bycatch—the nature and extent of discarded fish or wildlife accidentally caught when fishing for the target species.

These criteria are useful for companies to evaluate their supply. However, it is more common for firms to look to the ratings from the Monterey Bay Aquarium (*Seafood Watch)* and the Blue Ocean Institute (*Guide to Ocean Friendly Seafood)* that evaluate fish options against these criteria. The "red"-listed fish from these organizations are avoided and the "green"-listed choices are preferred. Among the favored choices that are also good for human health (e.g., low in environmental contamination such as mercury and high in long-chain omega-3 fatty acids), are albacore tuna (troll or pole caught from the United States or British Columbia), freshwater coho salmon (farmed in tank systems from the U.S.), farmed oysters, wild caught Pacific sardines, farmed rainbow trout, and wild caught salmon (from Alaska) (Monterey Bay Aquarium 2011).

Another approach to finding sustainable seafood options is selecting certified fish and seafood. The World Wildlife Fund (WWF) evaluated programs for wild caught fish and determined that the Marine Stewardship Council (MSC) was the leading certification program (WWF 2009).

An additional factor to evaluate is the form of transportation used to deliver the seafood from the waters to its destination. Air transport comes with a high environmental toll and should be avoided. The greenhouse gas emissions are estimated to be about 40 times greater for air freight than truck (Duchene 2009).

In the end, transparency from the boat making the catch through to the point of sale is critical to ensure that all the considerations are able to be addressed and properly managed.

## 3.2.1 Farmed fish

Aquaculture, or fish farming, produces half the seafood consumed globally and its role in our food supply will be increasing (FAO 2013a; National Geographic). While farmed fish avoids many of the problems with wild fishing, it does not avoid all issues (see Table 3.1). Farmed fish may use wild fish

**Table 3.1** Potential environmental issues of farmed fish

| Concern | Source |
|---|---|
| Pollution of inland and coastal waters | Release of water into the environment that includes excess feed, cleaning chemicals, antibiotics, feces |
| Ecosystem destruction | Farm construction |
| Feed needs contributing to challenges with wild caught fish (overfishing, ecosystem destruction) | Farmed species fed wild species and inefficient use of feed to grow certain species (e.g., shrimp, salmon) |
| Escaped farmed fish (escapes) that may endanger natural species and ecosystems | Offshore operations using cages and pens that insufficiently enclose fish |
| Genetically modified fish that escape and may endanger natural species and ecosystems | Not yet approved for use, but genetically modified salmon may be approved in the United States |
| Broodstock contributing to challenges with wild caught fish | Some aquaculture uses wild-caught fish to grow the stock |
| Fish illness and welfare | Unsanitary or overstocked conditions contribute to fish illness and welfare issues |

for feed, which needs to be caught and is subject to the same issues discussed earlier. Farmed salmon requires over 3 pounds of wild anchovies to produce 1 pound of salmon and farmed tuna requires over 15 pounds of feed per pound produced (Monterey Bay Aquarium). This feed-to-product ratio is very inefficient. A more sustainable farmed fish option would be one that does not require fish in the feed and that has a more efficient feed-to-product ratio. This includes tilapia, catfish, and shellfish (Monterey Bay Aquarium). Fish farms can be located in areas near native fish habitats. But this comes with environmental issues including the millions of farmed fish that escape (Monterey Bay Aquarium). The impact on the environment of these escapes is uncertain, but could threaten wild fish. Escapes can be avoided with equipment such as closed cages and management approaches. Fish farming also can have impacts similar to other farming, including habitat damage, pollution issues from waste and chemical use (e.g., pesticides and antibiotics), and disease. Aquaculture is growing rapidly, expected to double by 2030 (FAO 2013a). This growth will be concentrated in developing countries where effective laws need to be developed and enforced to avoid possible environmental issues (Baldwin 2014), so careful consideration even for farmed fish is necessary.

For farmed fish there are credible certification programs such as Best Aquaculture Practices Certification and the Environmental Defense Fund (for farmed shrimp and salmon) that can be leveraged to find sustainable options.

### 3.2.2 Approaches to address seafood

There has been significant activity in the food industry to move toward responsibly harvested and farmed seafood. Many processors, retailers, and food service organizations have active programs to source from preferred options. This is an area of food sustainability that is much farther ahead than most. Greenpeace evaluated such activity and identified the following best practices that support positive change for seafood sourcing (Greenpeace 2013a):

- Corporate policy with requirements and benchmarks to prevent the use of unsustainable seafood.
- Support for external initiatives and partnerships designed to promote positive change in the oceans (e.g., legislation).
- Transparency of the supply chain and sustainability practices for consumers and other interested parties.
- Avoidance of unsustainable seafood (no red list species).

Companies often procure fish certified as sustainable by leading programs such as MSC and look to avoid red list fish (Baldwin 2014). Purchasers can collaborate with organizations such as the Sustainable Fisheries Partnership or FishWise to work with fisheries not yet in compliance with certification programs in order to improve their practices and earn certification. This requires transparency in the supply chain to verify sources.

#### 3.2.2.1 American tuna

American Tuna sells responsibly-caught canned tuna in all the Whole Foods stores in the United States and other outlets including A&P, Pret A Manger, and Sysco (American Tuna). Since its founding in 2005, American Tuna has sourced tuna from the United States Pacific Northwest with full traceability to vessels that use pole and line harvesting of the albacore tuna. Pole and line fishing is done by baiting the tuna with sardines and anchovies thrown into the water and then catching the tuna with a fishing pole. The amount of bait needed is small because fish caught this way are smaller, such as the albacore. This catching method is considered one of the most responsible ways to harvest tuna since it avoids the common shortcomings with other practices, such as bycatch of sharks, turtles, and many other fish with purse seines or longlines. This tuna is certified through MSC. American Tuna puts all the catch to use, even traditional fish processing wastes are used by crab and eel fisherman and for other purposes such as fertilizer and supplements.

#### 3.2.2.2 Darden restaurants' Lobster farm

Darden Restaurants is the largest full-service restaurant company in the world. One of the company's flagship restaurants, Red Lobster, purchases an estimated $850 million of lobster and seafood each year (Gunther 2012). With the supplies of lobster at risk, Darden looked to aquaculture to help address the problem, however, lobster aquaculture did not exist. Darden has been working to develop the world's first commercial-scale lobster farm in Malaysia (Darden Restaurants 2012). Farmed lobster is expected to be available in 2017. During the start-up period, efforts are underway to conduct a two-year environmental impact assessment to ensure that the farm protects the environment. The farm is aiming for the most efficient feed ratio possible and to avoid negative impacts to the surrounding environment. Darden's farm is a contract model where independent farmers raise the lobsters. The company has committed to train and support the farmers so they can become successful over the long term and have also committed to buy the grown lobsters.

#### 3.2.2.3 Mars' seafood policy

Mars, Inc. manufacturers and sells packaged food for people and is also a leading producer of pet care products. Seafood is a source of protein for pet food, so Mars developed a sustainable seafood sourcing policy for its pet care business with the aim to "source all fish globally from sustainable sources by 2020" (Mars). In order to achieve this goal, Mars worked with its suppliers and stakeholders to develop programs aimed to:

- use fish from wild stocks that are not threatened or are responsibly farmed.
- replace all wild whole fish and fish fillets with fish by-products and responsibly-farmed seafood products.
- develop and use alternatives to marine fish ingredients.

These programs include sourcing from MSC-certified seafood and following the sourcing recommendations from the Monterey Bay Aquarium Seafood Watch program.

## 3.3 Environmental impacts from livestock production

Animal-based food products and ingredients have significantly greater environmental impacts than other food products. Energy inputs per unit of food energy produced from livestock production is about 12 times greater than other agricultural products (Horrigan et al. 2002). The FAO determined that 14.5% of all human-induced greenhouse gas (GHG) emissions come from livestock production (FAO 2013b). The livestock sector is the largest user of land occupying about 78% of agricultural land (Bellarby et al. 2008; FAO 2013b). Pollution from livestock production can be significant, with manure generation estimated to be three times the amount of human waste produced (Pew Charitable Trusts 2008). Further, livestock manure is often contaminated with pesticides, heavy metals, excess nutrients, antibitoics, and growth hormones that can end up in the environment. Yet, the environmental toll from livestock production is expected to increase as more and more people in the world move to eating diets abundant in animal protein.

### 3.3.1 Greenhouse gas emissions

No other part of the food supply chain has a larger carbon footprint than the livestock sector (FAO 2013c). This is due to three main sources of GHGs: (1) digestive by-products of the animal (enteric emissions), (2) manure, and (3) feed production (see Figure 3.6). This is not to overlook other sources of GHGs such as energy use, since these also add up. The large GHG impact is due to the types of gases produced from livestock production, especially methane.

Methane is 25 times more potent a GHG than carbon dioxide. Livestock production is responsible for 60% of global methane emissions and this is the dominant GHG from livestock, representing 44% of emissions (the rest being

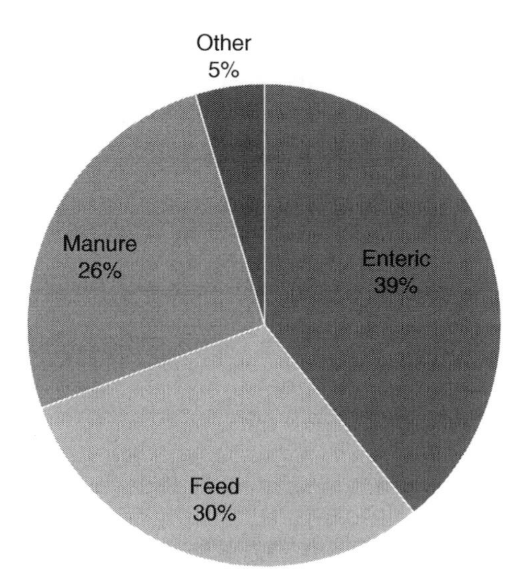


**Figure 3.6** GHG emissions from livestock (adapted from FAO 2013).

equally from carbon dioxide and nitrous oxide, in terms of carbon dioxide equivalents) (Bellarby et al. 2008; FAO 2013b).

Certain types of animals, called ruminant animals, digest food differently than humans using a four-part stomach. Many food products come from ruminant animals such as cows, sheep, goats, bison, and deer. This unique digestion process allows the animal to consume plants that are normally indigestible such as grasses and other forages. One of the by-products of their digestion is methane. Methane produced during digestion is mostly expelled through belching and, to a lesser extent, flatulence. This digestion by-product, enteric emissions, makes ruminants the largest contributors to the GHG impact from livestock (FAO 2013b). Nonruminants also produce methane but much lower levels are released into the atmosphere. Enteric methane is responsible for about 40% of the livestock emissions (FAO 2013b). Beef and cow milk production contribute 41 and 20% of the sector's emissions, respectively (FAO 2013b). Pig meat and poultry meat/eggs contribute a much lower proportion of the total livestock emissions each contributing 9% and 8%, respectively (FAO 2013b).

Ruminant animals' manure is another source of methane, along with the GHG nitrous oxide. GHGs from the manure of all types of animals, however, has become a significant issue due to the production method of concentrating livestock. Concentrated animal feeding operations (CAFOs) typically house a thousand or more animals in confinement (also leading to concerns about animal health and welfare issues). Instead of the manure naturally being spread over pastureland during grazing, the manure is collected and highly concentrated. This leads to significant waste in one area. It has been estimated that a dairy farm with 2,500 cattle produces as much waste as over

400,000 people. The state of North Carolina has so many CAFOs for hog production that the state produces as much waste from hogs as all the people living in North Carolina, California, New York, Texas, Pennsylvania, New Hampshire, and North Dakota combined (Lappé 2010).

There is currently no practical way of treating this concentration of manure or trying to recycle it back into nature. Consequently, CAFOs commonly manage this amount of waste by pooling it and allowing it to slowly decompose. This process releases significant amounts of methane. The release of methane from manure is unique to this means of handling; manure left in the pasture releases little-to-no methane (EPA 2010). Thus, different manure handling, storage, and processing approaches (e.g., through different production techniques or composting or anaerobic digestion of collected manure) can help reduce these emissions which are about a quarter of all livestock's emissions (FAO 2013b).

The production of feed for livestock is another leading source of GHG emissions in this sector. Nitrous oxide emissions come from the fertilization of crops where the applied nitrogen escapes into the atmosphere (with both organic and synthetic fertilizer). The production of fertilizer also causes GHG emissions, both nitrous oxide and carbon dioxide. Converting natural habitats to produce feed is yet another source of released carbon dioxide. The contribution of GHGs from the combination of these sources adds up to about 30% of livestock emissions (FAO 2013b).

There are already proven solutions to help reduce livestock-based emissions but they are not yet widely used (FAO 2013b). The FAO estimates that a 30% reduction of GHG emissions would be possible if producers adopted the currently available best practices for their region (FAO 2013b). Some of the recommended approaches include better quality feed and feed balancing to lower enteric and manure emissions, improved breeding and animal health to help shrink the herd overhead, manure management practices that ensure the recovery and recycling of nutrients and energy contained in manure, and improvements in energy use efficiency (FAO 2013b). Land management approaches (e.g., grazing) on the farm can also create a carbon sink to offset GHG emissions. Livestock feed impacts can be reduced with more sustainable options such as pasture feeding or organic feed (Cederberg 2003). For example, it has been shown that beef production on organic pastures requires half as much energy as conventional practices (Pimentel, Williamson, Alexander, Gonzalez-Pagan, Kontak, & Mulkey 2008).

The FAO notes that approaches to reducing GHG emission can provide both environmental and economic benefits since the GHG emissions that are concerning to the environment also represent inefficiencies in the system that waste money (FAO 2013b). This is the case across the world, with most opportunities for improvement in developing countries. While areas such as North America and Europe have lower GHG emissions per product, the production volume is large in these areas so every bit of improvement is significant. This is not to overlook the potential for reducing the overall reliance on/consumption of livestock products as well as efforts to reduce wasted food through the supply chain as important factors to addressing the issues (Bellarby et al. 2008).

### 3.3.2 Land use and pollution from livestock production

Livestock production is the largest user of land, using up a third of all land on the planet (FAO 2006). This amounts to 70% of the land used for agriculture while livestock only represents 40% of the agricultural gross domestic product (FAO 2006). This heavy demand for land has been a key contributor to global deforestation, especially of high conservation value lands.

The land requirements come from the space to raise the animal as well as the land needed to produce the feed for the animal. Land requirements are lower for milk and eggs, followed by pork and chicken meat, with beef requiring the most land (see Table 3.2) (de Vries & de Boer 2010).

The issues with this heavy land demand include GHG emissions, land degradation, limited freshwater availability, and biodiversity loss, among others. GHG emissions with land use arise through conversion of natural habitats to agriculture since current agricultural land management practices do not sequester carbon effectively (as natural habitats do). Land compaction and erosion from livestock production reduce the land's productivity potential and value. In the United States, livestock is responsible for 55% of soil erosion and sediment (FAO 2006). This is estimated to cost $1.2 billion per year (FAO 2006). Both land conversion and land degradation alter water cycles which can reduce freshwater availability (compounding the significant use of water for livestock production, 8% of total water) and is responsible for much of the global biodiversity loss that is altering the availability of food and ecosystem services (e.g., nutrient recycling, pollination, climate control, etc.) (FAO 2006).

The livestock sector is estimated to be the largest source of water pollution (FAO 2006). Manure is a leading cause of pollution along with feed production and land degradation. Manure traditionally has been a resource to recycle nutrients into the land, however, the current system is not balanced and concentrated amounts of manure are generated. This manure contains excess nutrients that pollute water and cause human health problems and eutrophication,

**Table 3.2** Land use requirements to produce livestock products (de Vries and de Boer 2010)

| | $m^2$/kg of Product | $m^2$/kg Protein | Average Daily Intake, $m^2$ |
|---|---|---|---|
| **Milk** | 1.1–2.0 | 33–59 | 0.62–1.1 |
| **Eggs** | 4.5–6.2 | 35–48 | 0.16–0.22 |
| **Chicken** | 8.1–9.9 | 42–52 | 0.6–0.73 |
| **Pork** | 8.9–12.1 | 47–64 | 0.73–0.99 |
| **Beef** | 27–49 | 144–258 | 1.65–2.96 |

alterations in oxygen availability in the water that can lead to dead zones. About 70% of nitrogen fed to livestock ends up being excreted, which can then end up in the waterways if not managed properly (FAO 2006). Manure also carries antibiotic and growth-enhancer residues that were fed to the animal. This can contribute to the concern with antibiotic resistance discussed earlier in the chapter. In developed countries where there are regulations aimed to prevent manure from polluting the water and environment, there are still issues with regulatory loopholes or regular mismanagement/spills. Runoff is also typically a source of pollution. These causes have contributed 32% of the nitrogen load and 33% of the phosphorus load into freshwater resources in the United States (FAO 2006).

The improvement opportunities discussed to address GHG emissions from livestock production can help address the issues with land use and pollution since they can improve the efficiency of the system. Improving feed to have better feed conversion and nutrient utilization are improvement opportunities, for example. In addition, there are best practices in manure and land management that can not only avoid the problems with livestock production, but also contribute positively to the land. Manure application to the land for fertilizer can be an effective way to reduce pollution (FAO 2006). When appropriately used, it recycles the nutrients into the soil and also improves the organic structure of the soil, which helps the soil's ability to retain nutrients and water. Well-managed grazing land (e.g., avoid overgrazing and undergrazing) that is rich in biodiversity can help reduce the need for extra inputs such as pesticides and reduce impacts including erosion. In fact, healthy grazing lands can contribute to improving water quality, sequestering carbon, and provide food from marginal lands that otherwise may be unproductive (U.S. Department of Agriculture).

### 3.3.3 Approaches to address environmental impacts from livestock

The scale of the environmental challenges with livestock demand widespread action and collaboration. This level of activity is not yet a reality. Unlike animal welfare and seafood sustainability, environmental responsibility does not have clear benchmarks or best practices outlined. "Organic" production has served as a framework for environmentally responsible practices. The term "organic," however, can mean different things across the globe. Typically it refers to the avoidance of synthetic inputs such as fertilizer, pesticides, or growth enhancers (e.g., antibiotics or hormones) to produce food. The largest market for organic foods is the United States, which is run by through U.S. Department of Agriculture National Organic Program (Sahota et al. 2008). Organic meat and dairy are among the products with the fastest growth (Sahota et al. 2008).

Beyond organic certification, expectations for environmentally-responsible livestock production addresses (Greenpeace 2013b) the following:

- Climate through ecological practices with production efficiencies, effective manure management, and optimizing sequestration on the land.
- Land through judicious use with efficiencies in production and effectively using feed sources for animals (e.g., grass for ruminants, agricultural/food waste for monogastrics).
- Pollution avoided through optimizing the available nutrients in manure and careful management from its production to use.
- Biodiversity protected and maximized on the land and with the animals in production.
- Water conserved through the practices aimed to address the other issues.

#### *3.3.3.1 Ben & Jerry's caring dairy*

Ben & Jerry's is a global ice cream brand owned by Unilever. In its early days when the company started as an independent brand, responsibility was embedded into everyday actions. Over the years, responsibility took many forms from supporting the local community to seeking out ingredients that were sustainably produced (Cohen & Greenfield 1998). A more recent development in the company's commitment to sustainability is a program called Caring Dairy. This is a voluntary program with its milk producer cooperatives that aims for improvement on eleven sustainability indictors on the farm (Ben & Jerry's b). These include areas such as soil fertility, soil loss, nutrient management, biodiversity, water, and pest control, among others including animal welfare, to facilitate reduced environmental impacts, improve quality and stable supply, and enable cost savings (see Table 3.3) (Ben & Jerry's a, b). In the area of nutrient management, for example, manure is recommended as a fertilizer with guidance on timing for application and other practices to optimize its use (Ben & Jerry's b). Soil loss management includes guidance on measuring organic matter in soil, use of cover crops, and rotational grazing to avoid water runoff and soil erosion and to encourage crop regrowth without overgrazing (Ben & Jerry's b).

The program engages with the producers by first assessing aspects of the dairy farm operation to determine areas of improvement. Educational opportunities and assistance are provided to the producer to help them implement the changes. In addition, the company provides financial incentives for participating farmers. The approach aims to have "happy farmers, happy cows, and a happy planet."

**Table 3.3** Ben & Jerry's caring dairy criteria (Ben & Jerry's a)

| | |
|---|---|
| **Soil health** | • Soil organic matter management<br>• Crop rotations<br>• Soil health monitoring<br>• Fertilizer and manure application<br>• Soils and farm mapping<br>• Record keeping and analysis<br>• Animal management |
| **Soil loss management** | • Soil Cover: User of cover crops, buffer strips, and perennials<br>• Tillage practices managing pasture<br>• Crop rotation and use of perennial crops in rotation<br>• Surface water protection<br>• Managing storm water and runoff<br>• Erosion management<br>• Monitoring topsoil |
| **Nutrients** | • Nutrient management record keeping<br>• Application of manure<br>• Applications of fertilizer<br>• Timing of applications<br>• Soil testing<br>• Equipment calibration<br>• Dietary phosphorus supplement management<br>• Continuing education |
| **Farm financials** | • Use and maintenance of a business plan<br>• Succession and retirement planning<br>• Understanding and utilizing financial statements<br>• Financial stability<br>• Debt management |
| **Social human capital** | • Community involvement<br>• Work/life balance<br>• Fair and legal treatment of all employees, and participation in labor training programs<br>• Responsible youth labor management<br>• Employee care<br>• Health insurance for all family members<br>• Farm safety and training |
| **Pest management** | • Pest identification and threshold monitoring<br>• Pesticide selection<br>• Pesticide applications and timing<br>• Record keeping<br>• Fly control<br>• Weed control<br>• Use of biological controls |
| **Biodiversity** | • Genetic diversity of crops<br>• Natural area conservation<br>• Management of riparian areas<br>• Pasture and crop management<br>• GMOs |

**Table 3.3** Ben & Jerry's caring dairy criteria (Ben & Jerry's a) (*continued*)

| | |
|---|---|
| **Animal husbandry and welfare** | • Use of herd health plan, monitoring, and protocols<br>• Calves<br>• Animal husbandry and safety training<br>• Cow nutrition<br>• Animal monitoring<br>• Locomotion<br>• Sanitation<br>• Body condition scores and hock lesions<br>• Milk quality<br>• Cull rate<br>• Animal environment<br>• Facilities<br>• Handling and transport<br>• Special-needs animals |
| **Energy** | • Use of energy efficient equipment: lighting, pumps, compressor, water heating, ventilation<br>• Equipment maintenance<br>• Other energy conservation measures<br>• Renewable energy technologies<br>• Greenhouse gas emissions |
| **Water** | • Manage runoff from barnyard and farm buildings<br>• Manure and fertilizer storage<br>• Silage leachate capture<br>• Milk house waste management<br>• Protecting on-farm surface water<br>• Water use plan<br>• Water use management |
| **Local economy** | • Support for local businesses and charities<br>• Farm management to enhance the working landscape<br>• Recreational access<br>• Waste management and recycling, reduction, rethink<br>• Air quality<br>• Adopting new practices<br>• Fair and meaningful compensation for employees |
| **Farm metrics** | • Electricity use per unit of milk produced<br>• Fuel use per unit of milk produced<br>• Mastitis cases per cow<br>• Fertilizer use per acre<br>• Milk Urea Nitrogen (MUN) testing score<br>• Percent of annually tilled acres cover cropped |

### *3.3.3.2 Niman Ranch*

Niman Ranch is a United States network of over 700 family farmers and ranchers that raise their livestock with high environmentally sustainable practices and animal welfare standards. Niman Ranch farms produce beef, pork, eggs, chicken, and lamb products. Niman Ranch approaches sustainability by requiring producers to "embrace the land" with approaches that protect resources.

Niman Ranch supports organic production principles but is not certified. This is because the company has found that there is not enough certified feed in the United States to support the production volumes of its farmers and ranchers without requiring importation of feed. Instead, Niman Ranch focuses on raising fewer animals on each farm to allow for responsible practices from manure management to grazing management. All products are raised without antibiotics or growth hormones and the following best practices (Niman Ranch):

- Mitigate soil erosion and/or loss through: maintaining pasture with coverage for livestock, crop rotation, rotational grazing, and responsible waste/manure management.
- Prohibit the use of concentrated liquid manure systems.
- Utilize buffer strips and grassed waterways.
- Promote agricultural biodiversity by using breeds which are uniquely suited for their specific environment.
- Practice genetic diversity to keep breeds healthy over generations.
- Maintain conservative livestock density stocking rates to improve the land resource over time.
- Raise livestock in geographies where feed sources are locally available to reduce the environmental impact of feed transport.
- Pay farmers a premium in accordance to our strict raising protocols.
- Establish a floor price for our farmers tied to the cost of inputs of feed and fuel.
- Provide a robust and growing marketplace for their livestock.

Niman Ranch consulted with experts from academia and nonprofits to help shape their approach to sustainability. The program has evolved over time and aims to continue to improve as more is learned. In addition to these environmental practices, Niman Ranch farmers follow strict animal welfare protocols based on recommendations from leading authority Dr. Temple Grandin. This includes the animals being raised outdoors or in deeply bedded pens where the animals have access to fresh, clean water and they can express their natural behaviors.

The commitment of each Niman Ranch producer is officially acknowledged with an affidavit to agree to implementation of the comprehensive program. The company provides full validation by verifying that producers follow their environmental and welfare requirements with thorough record evaluation and on-site inspections. Niman Ranch's leading approach provides transparency and assurance that their products are produced responsibly.

## 3.4 Summary

The overall environmental footprint from livestock and seafood production is significant but can be reduced. Livestock production has an important role in a more sustainable food production system but changes to the current system are needed. The term "ecological livestock" has been used to refer to this concept of a positive contribution to the system, replenish freshwater, revitalize biodiversity, build soil fertility, and restore land while also serving as a carbon sink (Greenpeace 2013a). To get to this vision, livestock will need to be raised more on grasslands, pasture, and residues for feed to minimize the use of arable land and protect natural ecosystems (Greenpeace 2013b). Treatment of livestock must be performed to high welfare standards to ensure their health and productivity and for the system to thrive. Companies engaged at this level have reported cost savings and a more secure supply. In order to reach these targets, the supply chain needs to be engaged all the way to the farm. Seafood sourcing is a leading example already improving critical ecosystems and fisheries. MSC reported that by 2013, 13 fisheries have completed stock improvements to reach best practice (or sustainable) levels and 22 fisheries have completed habitat and ecosystem improvement (MSC 2013). More efforts, such as seafood, where the industry engages up to the source of production can shift livestock from being the dominant source of sustainability challenges to one that delivers notable benefits (Table 3.4).

**Table 3.4** Summary of the principles–practices–potential of sustainability for animal and fish production and harvesting

| | |
|---|---|
| **Principle** | *Use of animals, fish, and seafood in the food supply optimizes their well-being and adds to environmental health.* |
| **Practices** | • Implement responsible animal and fish production techniques such as:<br>  • Well-managed grazing/feeding, land stewardship and stocking density, and effective manure management approaches (including responsible production techniques noted in no. 2)<br>  • Diverse breed production for health, resilience, and productivity<br>  • Productive rearing that respects the five freedoms of responsible animal care (freedom from hunger/thirst, discomfort, pain/injury/disease, inability to express normal behavior, and fear/distress)<br>  • Responsible seafood production and harvesting in the wild and in aquaculture (including responsible production techniques noted in no. 2)<br>• Source and support responsibly produced and harvested food |
| **Potential** | • Replenish water<br>• Revitalize biodiversity<br>• Build soil fertility<br>• Restore land and provide a carbon sink |

# Resources

Blue Ocean Institute's *Guide to Ocean Friendly Seafood* http://blueocean.org/programs/sustainable-seafood-program/ocean-friendly-substitutes/
Business Benchmark on Farm Animal Welfare: http://www.bbfaw.com/
Compassion in World Farming www.ciwf.org
Fish Watch: www.fishwatch.gov
Five Freedoms: http://www.fawc.org.uk/freedoms.htm
Monterey Bay Aquarium Seafood Watch http://www.montereybayaquarium.org/cr/seafoodwatch.aspx
Temple Grandin: http://www.grandin.com/
United Nations Food and Agriculture Organization www.fao.org

# References


American Humane Association. (2013). *Humane heartland: farm animal welfare survey*. Retrieved from http://www.americanhumane.org/assets/humane-assets/humane-heartland-farm-animals-survey-results.pdf

American Tuna. *About us*. Retrieved from http://www.americantuna.com/html/about.html

Amos, N., & Sullivan, R. (2012). *The Business Benchmark on Farm Animal Welfare 2012 Report*. Retrieved from http://www.bbfaw.com/wp-content/uploads/2010/08/BBFAW_Report_Low_Res.pdf

Baldwin, C. (2013). *Green eggs and ham: sustainability pillars for animal-based food production*. Retrieved from http://blog.purestrategies.com/blog/bid/272126/Green-Eggs-and-Ham-Sustainability-Pillars-for-Animal-Based-Food-Production

Baldwin, C. (2014). *Sustainable seafood: one fish, two fish, red fish, green fish*. Retrieved from http://blog.purestrategies.com/blog/bid/336632/Sustainable-Seafood-One-Fish-Two-Fish-Red-Fish-Green-Fish

Baldwin, C., & Atwood, R. (2014). *Pure strategies webinar: supply chain engagement for sustainable change*. Sustainability Management Association, January 23, 2014.

Bellarby, J., Foereid, B., Hastings, A., Smith, P., et al. (2008). *Cool farming: climate impacts of agriculture and mitigation potential*. Retrieved from http://www.greenpeace.org/international/Global/international/planet-2/report/2008/1/cool-farming-full-report.pdf

Ben & Jerry's a. *Caring dairy*. Retrieved from http://www.benjerry.com/values/how-we-do-business/caring-dairy#8timeline

Ben & Jerry's b. *The path to enlightenment*. Retrieved from http://www.benjerry.com/activism/inside-the-pint/dairy#.UkehwSR6NxF

Cederberg, C. (2003). Life-cycle assessment of animal products in environmentally friendly food processing. B. Mattsson and U. Sonesson (Eds.). Cambridge, England: Woodhead Publishing Limited.

Centers for Disease Control and Prevention. (2013). *Antibiotic resistance threats in the United States, 2013*. Retrieved from http://www.cdc.gov/drugresistance/threat-report-2013/pdf/ar-threats-2013-508.pdf#page=31

Charles, D. (2013). *Pig manure reveals more reason to worry about antibiotics*. Retrieved from http://www.npr.org/blogs/thesalt/2013/02/11/171690001/pig-manure-reveals-more-reason-to-worry-about-antibiotics?ft=1&f=139941248

Cohen, B., & Greenfield, J. (1998). *Ben & Jerry's double dip: how to run a values-led business and make money, too*. New York: Simon & Schuster.

Compassion for World Farming. *Tesco's improved welfare chicken*. Retrieved from http://www.compassioninfoodbusiness.com/case-studies/tesco%E2%80%99s-improved-welfare-chicken/

Convention on Biological Diversity (CBD). (2010). Global biodiversity outlook 3. Retrieved from http://www.cbd.int/doc/publications/gbo/gbo3-final-en.pdf.

Darden Restaurants. (2012). Protecting the animal that built our company. Retrieved from http://www.darden.com/sustainability/default.aspx?lang=en&page=sustainability§ion=blog&mode=post&id=805

de Vries, M., & de Boer, I. J. M. (2010). Comparing environmental impacts for livestock products: a review of life-cycle assessments. *Livestock Science,* 128, 1–11.

Duchene, L. (2009). *Air freight doesn't always fly with sustainable seafood*. Retrieved from http://www.seafoodbusiness.com/articledetail.aspx?id=4294994142

Food and Agriculture Organization (FAO) of the United Nations. (2004). *The future of agriculture depends on biodiversity*. Press release October 15, 2004. Retrieved from http://www.fao.org/newsroom/en/focus/2004/51102/index.html

Food and Agriculture Organization of the United Nations. (2006). *Livestock's long shadow: environmental issues and options*. Retrieved from http://www.fao.org/docrep/010/a0701e/a0701e00.HTM

Food and Agriculture Organization of the United Nations. (2013a). *Fish to 2013: prospects for fisheries and aquaculture*. Retrieved from http://www.fao.org/docrep/019/i3640e/i3640e.pdf?utm_source=FCRN+Mailing&utm_campaign=966379807b-RSS_*|RSSFEED%3ADATE%3Aj+F+Y|*&utm_medium=email&utm_term=0_a29d7fdc4d-966379807b-297125489

Food and Agriculture Organization of the United Nations. (2013b). Tackling climate change through livestock: a global assessment of emissions and mitigation opportunities. Retrieved from http://www.fao.org/docrep/018/i3437e/i3437e.pdf

Food and Agriculture Organization of the United Nations. (2013c). FAO statistical yearbook 2013: world food and agriculture. Retrieved from http://www.fao.org/docrep/018/i3107e/i3107e00.htm

Forini, K., et al. (2005). *Resistant bugs and antibiotic drugs: state and county estimates of antibiotics in agricultural feed and animal waste*. Retrieved from www.edf.org/health/report/resistant-bugs-and-anitbiotic-drugs

Grandin, T. *Reply to the paper by Stan Curtis on animal state of being and welfare*. Retrieved from http://www.grandin.com/welfare/reply.to.stan.curtis.animal.welfare.html

Greenpeace. (2013a). Carting away the oceans: 7. Retrieved from http://www.greenpeace.org/usa/Global/usa/planet3/PDFs/oceans/CATO%20VII.pdf

Greenpeace (2013b). *Ecological livestock: options for reducing livestock production and consumption to fit within ecological limits, with a focus on Europe*. Retrieved from http://www.greenpeace.org/international/Global/international/publications/agriculture/2013/Ecological-Livestock.pdf

Gunther, M. (2012). *Darden is farming lobsters for the long-term*. Retrieved from http://www.greenbiz.com/blog/2012/08/14/how-darden-farming-lobsters

Horrigan L., et al. (2002). How sustainable agriculture can address the environmental and human health harms of industrial agriculture. *Environmental Health Perspectives,* 110, 445–456.

Humane Society of the United States. (2013). *Timeline of major farm animal protection advancements*. Retrieved from http://www.humanesociety.org/issues/confinement_farm/timelines/timeline_farm_animal_protection.html

Lappé, A. (2010). *Diet for a hot planet: The climate crisis at the end of your fork and what you can do about it.* New York: Bloomsbury.

Marine Stewardship Council (MSC). (2013). Global impacts: A summary. Retrieved from http://www.msc.org/business-support/global-impacts

Mars. *Fish*. Retrieved from http://www.mars.com/global/about-mars/mars-pia/our-supply-chain/fish.aspx

Monterey Bay Aquarium. *Aquaculture issue: Use of wild fish a bountiful harvest still depends on the sea*. Retrieved from http://www.montereybayaquarium.org/cr/cr_seafoodwatch/issues/aquaculture_wildfish.aspx

Monterey Bay Aquarium. (2011). *Turning the tide: The state of seafood*. Retrieved from http://www.montereybayaquarium.org/cr/cr_seafoodwatch/content/media/MBA_SeafoodWatch_StateofSeafoodReport.pdf

National Geographic. *Sustainable seafood*. Retrieved from http://ocean.nationalgeographic.com/ocean/take-action/sustainable-seafood/

Niman Ranch. *Defining sustainable agriculture*. Retrieved from http://www.nimanranch.com/environmentally_sustainable.aspx

Pew Charitable Trusts. *Ending illegal fishing project*. Retrieved from http://www.pewenvironment.org/campaigns/ending-illegal-fishing-project/id/8589941944

Pew Charitable Trusts. (2008). *Putting meat on the table: Industrial farm animal production in America*. Retrieved from http://www.pewtrusts.org/our_work_report_detail.aspx?id=38442

Pew Charitable Trusts. (2013). *Record-high antibiotic sales for meat and poultry production*. Retrieved from http://www.pewhealth.org/other-resource/record-high-antibiotic-sales-for-meat-and-poultry-production-85899449119

Pimentel, D., Williamson, S., Alexander, C. E., Gonzalez-Pagan, O., Kontak, C., & Mulkey, S. E. (2008). Reducing energy inputs in the U.S. food system. *Human Ecology*, 36, 459–471. doi: 10.1007/s10745-008-9184-3

Reed, K. (2011). *Farm animal welfare: The business case for action*. Retrieved from http://www.compassioninfoodbusiness.com/wp-content/uploads/2011/11/Farm-Animal-Welfare-The-Business-Case-for-Action.pdf

Sahota, A., et al. (2008). Ecolabeling and consumer interest in sustainable products. In Cheryl Baldwin (Ed.), *Sustainability in the food industry* (pp. 159–184). Ames, IA: Wiley-Blackwell.

Tesco. *Tesco farming*. Retrieved from http://realfood.tesco.com/our-food/tesco-farming.html

Unilever. (2012). *Unilever sustainable living plan: Progress report 2012*. Retrieved from http://www.unilever.com/images/USLP-Progress-Report-2012-FI_tcm13-352007.pdf

U.S. Department of Agriculture. *Sustainable grazing lands: Providing a healthy environment*. Retrieved from http://www.nrcs.usda.gov/Internet/FSE_DOCUMENTS/stelprdb1043496.pdf

U.S. Environmental Protection Agency (2010) Emissions inventory 2010. Inventory of U.S. greenhouse gas emissions and sinks: 1990-2008. Retrieved from http://www.epa.gov/climatechange/ghgemissions/usinventoryreport.html

U.S. Environmental Protection Agency. (2010). *Emissions inventory 2010. Inventory of US greenhouse gas emissions and sinks: 1990–2008*. Retrieved from http://www.epa.gov/climatechange/emissions/usinventoryreport.html

U.S. Food and Drug Administration. (2013). *FDA annual report on antimicrobials sold or distributed for food-producing animals in 2011*. Retrieved from http://www.fda.gov/AnimalVeterinary/NewsEvents/CVMUpdates/ucm338178.htm

World Economic Forum. (2013). *The dangers of hubris on human health*. Retrieved from http://reports.weforum.org/global-risks-2013/risk-case-1/the-dangers-of-hubris-on-human-health/

World Health Organization. (2012). *The evolving threat of antimicrobial resistance: Options for action*. Retrieved from http://www.who.int/patientsafety/implementation/amr/publication/en/index.html

World Wildlife Fund (WWF). *Facts about our earth.* Retrieved from http://wwf.panda.org/about_our_earth/

World Wildlife Fund. (2009). *Assessment of on-pack, wild-capture seafood sustainability certification programmes and seafood ecolables*. Retrieved from http://awsassets.panda.org/downloads/full_report_wwf_ecolabel_study_lowres.pdf

# 4 Processing

*Principle: Food and ingredient processing generates resources and requires minimal additional inputs and outputs.*

Food ingredients and products are nearly all processed through some means. This includes heating, washing, separating, cutting, mixing, or other processing method or combinations of methods. As a result, food processing can be a notable source of environmental issues across the life cycle of food and beverage products. Environmental harm comes from resource use such as energy and water, and the generation of waste. Energy costs are projected to increase, and availability to consistent freshwater will decrease. As a result, efforts to address the environmental concerns provide value from cost savings to risk reduction, and because of this, companies have been progressing their operational efficiency with substantial progress.

## 4.1 Energy

Food processing is a dominant user of energy in the food supply chain. In the United States processing uses close to 20% of the total energy in the food cycle, second only to consumer energy use for storage and preparation (see Figure 4.1) (Canning, Charles, Huang, Polenske, & Waters. 2010). Energy demand for processing has been increasing (Canning et al. 2010). This upsurge has been greater than any other part of the supply chain (Canning et al. 2010). A key driver is the shift to more processed and packaged food (e.g., packaged salad).

### 4.1.1 Energy sources and impacts

The type of energy consumed determines the environmental burden from energy use. This is because different energy sources affect the environment differently. Fossil, nuclear, and renewable sources are generally the main

*The 10 Principles of Food Industry Sustainability*, First Edition. Cheryl J. Baldwin.

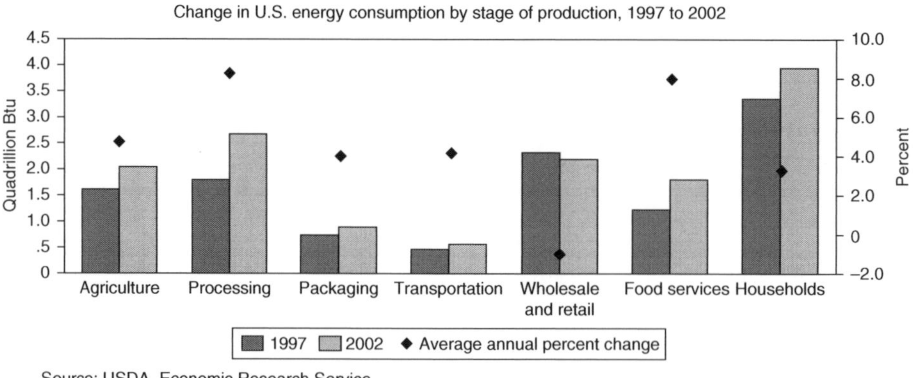


**Figure 4.1** Energy use across the food supply chain (Canning et al. 2010).

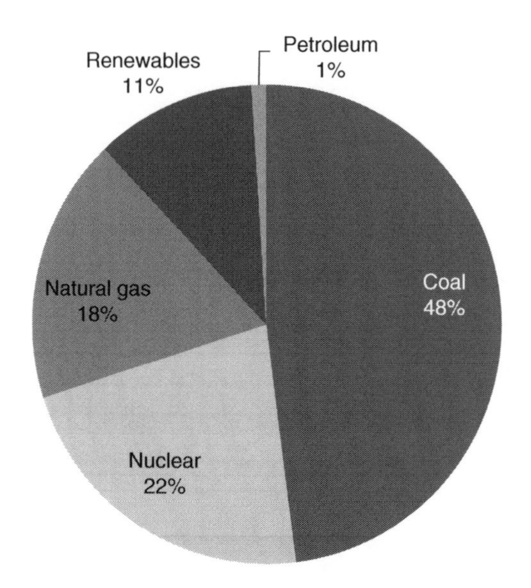


**Figure 4.2** Sources of energy in the United States (United States Energy Information Administration, United States Department of Energy 2010).

sources of energy. Of these, the dominant energy sources in the United States are fossil sources, coal and natural gas and nuclear energy (see Figure 4.2). Renewable energy sources—chiefly solar, wind, and geothermal—are minor contributors. These renewables and other alternative energy options are growing in use in food processing (Canning et al. 2010). Some alternative sources include combined heat and power, fuel cells, and digesters. The growth of renewable and alternative energy is a beneficial trend since they typically have the least impact on the environment. These sources use little to no water and emit very little or no greenhouse gases (GHGs). GHG emissions are a leading environmental challenge with energy use. The reason is that these gases (carbon dioxide, methane, nitrogen oxides, etc.) are emitted during the generation of energy and contribute to climate change.

Fossil fuel energy sources are derived from millions of years old buried plants and animals that have been exposed to intense heat and pressure over time. The energy that these fossilized plants and animals originally obtained from the sun is concentrated and released when we burn them. While fossil fuels come from nature, they are not renewable energy sources because they cannot be replenished in a human time frame. Depleting this limited resource is an irreversible loss; yet, there are additional drawbacks including emitting pollutants, toxins (e.g., carcinogens and neurotoxins), and GHGs as well as their energy generation demanding large amounts of water.

Coal is the primary energy source in the United States (see Figure 4.2) and emits more GHGs than any other energy source. While methods are being developed to reduce the GHG emissions from coal by capturing the carbon dioxide emitted and storing it underground or liquefying it for other uses (sometimes referred to as clean coal), none have been implemented. Coal has additional drawbacks from mining and transporting the fossil fuels to energy generation issues as outlined in Table 4.1.

**Table 4.1** Environmental drawbacks from coal energy use (Union of Concerned Scientists)

| | |
|---|---|
| **Mining and transporting** | • Damages the land from surface mining (two-thirds of coal is mined this way)<br>• Threatens the safety of miners<br>• Leaves behind toxic pollutants that contaminate the soil and water<br>• Requires significant transportation<br>• Contaminates the environment with toxins such as mercury and lead from coal dust all along the transportation route |
| **Energy generation** | • Requires significant amounts of water (2.2 billion gallons of water used for cooling the generation of 3.5 billion kilowatt-hours per year, enough to power a city of about 140,000 people)<br>• Contributes significantly to GHG emissions and climate change<br>• Emits pollutants that cause acid rain that damages forests, lakes, and buildings<br>• Emits small, airborne particles that can penetrate in the lungs to cause health problems<br>• Creates toxic waste that includes heavy metals such as mercury, lead, and arsenic that are poisonous to humans and wildlife<br>• Discharges waste heat to the environment, usually into the waterways, altering the ecosystem and causing such issues as decreased fish fertility<br>• Disposes water contaminated with chlorine and other toxins |

Natural gas has the lowest rate of GHG emissions of the fossil fuel energy sources (United State Environmental Protection Agency [EPA] 2009). Compared to the average air emissions from coal-fired generation, natural gas produces half as much carbon dioxide, less than a third as much nitrogen oxides, and 1% as much sulfur oxides at the power plant (EPA 2009). However, the mining and extraction of natural gas from the earth has its share of issues. Natural gas extraction from the earth, especially using hydraulic fracturing, or fracking, is emerging as a dangerous practice. Fracking is suspected of causing toxic water contamination from the chemicals used in fracking as well as from natural gas leakages (Schrope; Natural Resources Defense Council).

Nuclear energy is derived from uranium, a radioactive mineral mined from the earth. Use of uranium for energy depletes the nonrenewable resource and creates radioactive waste along the way. Radioactive material and waste are both harmful to humans (e.g., they cause cancer) and the environment (e.g., poisoning wildlife). While scientists achieved the first nuclear chain reaction in 1942, there is still no reliable way to store nuclear waste, some of which remains deadly for many thousands of years. Nuclear energy generation does not produce GHG emissions, unlike the burning of fossil fuels, which is why some favor the use of nuclear energy.

The GHG emissions that result from energy use are dependent on the source of the energy. The most emissions arise when coal is used and the least when nuclear or renewable resources are used. Each geographic region has its own mix of energy sources and ends up having a substantial difference in emissions. In the United States, the highest GHG emissions come from the Rocky Mountain region with 1,825 lb carbon dioxide equivalent/MWh and the lowest is in Upstate New York with 500 carbon dioxide equivalent/MWh (EPA 2011). This significant difference is due to the Rocky Mountain region relying far more on coal than Upstate New York which utilizes substantial amounts of hydroelectric and nuclear-based energy.

#### *4.1.1.1 Nonenergy GHG emissions*

There are GHG emissions from food processing beyond energy use. The most common sources are refrigerants for cooling, methane from on-site wastewater treatment, and carbon dioxide used directly in processing for various purposes. This is estimated to be about 10% of processing GHG emissions (EPA 2008). There are opportunities to address this smaller proportion of GHG emissions from processing. Many of these correspond to the methods to reduce energy-based GHGs, including maintenance and addressing leaks in refrigeration and carbon dioxide systems, and using refrigerants that do not have such a GHG impact when possible (e.g., ammonia). For wastewater treatment, the methane can be used as a source of energy and is a growing practice.

## 4.1.2 Energy use in food processing

From the electricity used to drive the motors to the natural gas that fuels heating, energy is a critical resource in food processing (see Table 4.2) (Morawicki 2012). Corn wet milling, beet sugar, soybean oil mills, malt beverages, meatpacking, canned fruits and vegetables, frozen fruits and vegetables, and baked goods are the leading energy users (EPA 2007). Heating and cooling are estimated to demand 75% of energy in food processing plants (see Figure 4.3) (EPA 2007). This is demonstrated with corn wet milling, the process used to make starch, sweeteners (e.g., high-fructose corn syrup), and ethanol from corn. It is the largest user of energy in food processing consuming 15% of the energy for all food processing, due to it being a very heat-intensive process (Galitsky, Worrell, & Ruth 2003). Another heat-intensive process is baking. Table 4.3 outlines typical finished food categories' energy demand showing that bakery products had the largest increases. These two processes illustrate

**Table 4.2** Energy uses in a food processing plant (Morawicki 2012)

| | |
|---|---|
| **Electricity** (typically coal or nuclear) | • Motors<br>• Refrigeration<br>• Air compressors<br>• Fans<br>• Process equipment<br>• Conveyors |
| **Process heat** (typically natural gas, oil, propane, or biomass) | • Steam boilers<br>• Dryers<br>• Ovens<br>• Fryers |

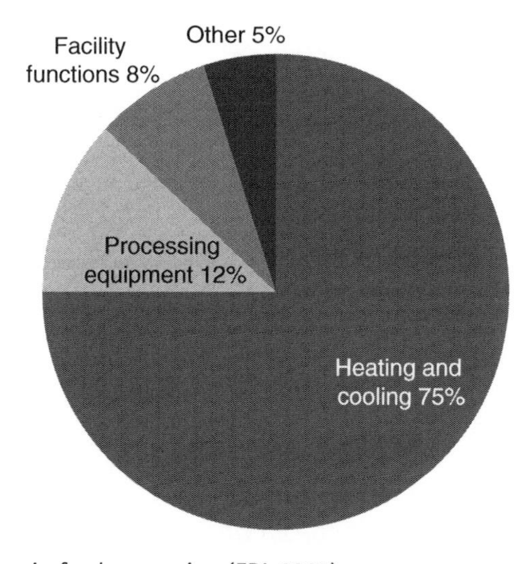


**Figure 4.3** Energy use in food processing (EPA 2007).

**Table 4.3** Per capita energy flows of a sample of expenditure categories (Canning et al. 2010)

| Food Product | Energy Use in 2002 (Btu) | Average Yearly Change from 1997–2002 (%) |
|---|---|---|
| Fresh vegetables | 25 | 11.7 |
| Fresh dairy | 251 | 10.7 |
| Beef | 360 | 8.8 |
| Cereal products | 468 | 10.4 |
| Poultry products | 585 | 7.9 |
| Beverages | 765 | 6.5 |
| Baking products | 1,129 | 13.0 |
| Snack, frozen, canned, and other foods, spices, and condiments | 1,422 | 9.6 |

opportunities to improve energy efficiency that will be discussed since they translate to other processes.

The several steps of drying and evaporation represent 80% of the energy use for corn wet milling (Galitsky et al. 2003). This is largely from fuel and steam. Electricity is used to drive pumps and motors. With such a high energy demand from both heat and electricity many facilities have begun to use site-produced energy namely from combined heat and power (Galitsky et al. 2003). This is where an energy source is used to generate electrical power and the heat released from that process is also used. This cogeneration of energy is nearly two times more efficient than traditional systems at a power plant or boiler (EPA). Minimizing added water in the process may also be an opportunity to reduce the energy demand since removing water through evaporation and drying processes have high energy requirements.

Similar to corn wet milling, the baking stage of producing goods such as rolls, cookies, crackers, cakes, and bread dominates energy needs (Masanet, Therkelsen, &, Worrell 2012). However, when frozen goods are produced the freezing step overtakes baking (Masanet et al. 2012). Additionally, the impact of cross-site energy users including air-compressors, boilers, and motors that together consume significant energy should not be overlooked. Energy efficiency measures would be best targeted at the high-energy users, starting with heating and cooling, followed by processing equipment, and facilities functions.

Heating process efficiency measures, for example, include cleaning and maintaining equipment (especially heat transfer surfaces and condensers), improving insulation, fixing steam leaks, minimizing heat loss in boiler blowdown water, returning condensate to boilers, and recovering heat from processes (Morawicki 2012). Similar opportunities are seen with other processes and equipment. Compressed air systems are notable because they are widely used and since they are an inefficient energy user that usually have efficiency opportunities (University of Minnesota a). Compressed air systems need regular maintenance, leaks reduced, to have unnecessary compressed air turned

**Table 4.4** The range of energy efficiency opportunities (Masanet et al. 2012)

| Minimal Investment ———————————→ Larger Investment | | | |
|---|---|---|---|
| Maintenance | Minor Adjustments | Process Adjustments | Efficient Technology |
| • Clean and tune equipment<br>• Fix leaks<br>• Turn off idle equipment and lighting | • Add insulation<br>• Add timers and controls to lighting and equipment<br>• Update lighting | • Reduce downtime<br>• Adjust ingredient handling (controls, temperature) | • Upgrade older equipment<br>• Generate heat and electricity on-site (e.g., combined heat and power, solar panels, digesters) |

off, the lowest possible pressure used, to be properly sized regulators and pipes, inlet air temperature reduced, or to be replaced with alternatives such as air blowers and fans to cool products (Masanet et al. 2012).

Table 4.4 outlines measures that can help address these opportunities (Masanet et al. 2012). Low cost options begin with straightforward maintenance approaches including keeping equipment clean (especially filters and function surfaces), properly adjusted, and free of leaks as well as using equipment only when it is needed (i.e., turn off idle equipment). A bakery in Brazil addressed the leaks in its compressed air system with a payback period of just nine months (Masanet et al. 2012). Minor adjustments to equipment and processes from installing low-cost controls to adding insulation can save energy. A bread plant in the United States had a payback of 10 months and a 25 MWh/year savings from installing occupancy sensors throughout the facility (Masanet et al. 2012). Changes to the process itself offer additional opportunities. These can include changes to batch timing or scheduling to reduce down times or modifying ingredient handling to reduce heating needs, such as tempering ingredients before processing. Finally, investing in energy-efficient equipment is an important opportunity that may seem daunting with its potential capital requirements, but payback can be significant. To illustrate, about 95% of a motor's cost over its lifetime is from the energy costs to run it, with only 5% coming from the initial cost of the purchase (Masanet et al. 2012).

Producing energy on-site is within reach for more companies than many realize. There are several options that are being adopted in the food industry, including combined heat and power systems, solar panels, and biomass digesters. Food processing plants can use biomass by-products for producing combined heat and power. For example, sugar mills use their bagasse residues (the ligno cellulosic material left after the sugar has been extracted from the cane) to produce energy through combined heat and power co-generation (Food and Agriculture Organization of the United Nations [FAO] 2011). Wet processing wastes such as tomato rejects and skins, and pulp wastes from juice processing can be used as feedstocks in anaerobic digesters to produce methane biogas. The methane can be used to generate heat or power or even as vehicle fuel after removing impurities (FAO 2011). The EPA featured two companies in a report that installed combined heat and power systems with payback of two and a half to four years, with one site saving $1.5 million each year (Masanet et al. 2012). On-site energy production may be worth a careful evaluation and may even come with rebates and incentives as illustrated in the Sierra Nevada example that follows.

### 4.1.3 Sierra Nevada's energy and climate program

Sierra Nevada is a premier craft brewer in the United States. They produce nearly 800,000 barrels of beer each year with a sustainability program that effectively addresses issues such as waste, energy, GHG emissions, and water. Part of this is a leading energy management program. The company aims for near energy self-sufficiency through the combination of on-site energy production and dramatic energy conservation measures. A partnership with their power company, Pacific Gas and Electric Company (PG&E), supported these efforts with technical assistance and rebates over the three decades of the company's progress.

Sierra Nevada's energy efforts are grounded in comprehensive conservation approaches. Past projects include upgrading lighting systems to leverage natural lighting, using lighting sensors and timers, retrofitting single-speed motors with variable-speed motors, upgrading and insulating boilers, upgrading compressor systems, installing energy monitoring systems to view real-time usage, and updating appliances and fixtures (PG&E 2011). Sierra Nevada also achieved efficiencies with heat recovery projects. Vapor condensers and heat exchangers were installed to capture and recycle heat from boilers, kettles, and fuel cells.

Particularly notable is the company's success with on-site energy production. One of the drivers for the company's focus was to ensure a reliable energy supply for their manufacturing since an interruption can result in weeks of lost manufacturing (PG&E 2011). Sierra Nevada decided to use solar power and fuel cells to produce their own energy. The company claims to house one of the largest privately owned solar arrays in the country (Sierra Nevada). This includes 10,500 panels on the brewery roof, covered parking lot, day care facility roof, and rail spur that produce about 20% of the company's energy needs (PG&E 2011). The installation was supported by power company rebates, state incentives, and federal tax credits (PG&E 2011). The fuel cells run on natural gas to produce about 40% of the company's energy needs. To improve the fuel cells' efficiency, heat recovery units were added that produce steam used in the brewing process (NFPA 2013). This installation was also supported with external incentives, covering about 70% of the total costs (NFPA 2013). The company has taken an additional step by using anaerobic digestion of brewery wastewater to recover methane. The biogas is used along with natural gas to fuel the company's boilers (PG&E 2011).

In addition to these achievements with energy conservation and on-site production with less polluting sources, Sierra Nevada invests in projects off-site that capture or reduce carbon dioxide emissions through their power company. This initiative has offset nearly 42,000 tons of carbon dioxide emissions (PG&E 2011). The company also captures and recycles all of the carbon dioxide used and produced in its processes (PG&E 2011). This helps avoid purchasing carbon dioxide and transporting it from 100 miles away (Sierra Nevada). The company also voluntarily quantifies and reports greenhouse gas emissions to the Climate Registry.

Sierra Nevada's multi-faceted approach to energy and climate management sets a high bar, but demonstrates that it can be cleared for other food companies to strive toward.

### 4.1.4 Heinz's energy effort in energy and greenhouse gas emissions

Heinz is a United States-based company that sells the iconic Tomato Ketchup and other retail and foodservice products. Part of a comprehensive sustainability program, the company is on track to meet its ten year goals to reduce energy and GHG emissions by 20% by 2015 (per unit metric ton of production) (Heinz 2011). Impressively, absolute emissions have also decreased due to the effectiveness of the energy efficiency program. The company has implemented energy conservation efforts that range from low investment solutions to line reengineering and new technology installations. Heinz has done this with a strong employee-driven program. However, Heinz also seeks external assistance to help identify energy efficiency opportunities. The company partners with engineering and sustainability consultants to identify and implement changes.

Lighting design changes using daylight helped cut one site's electricity consumption by 50% (Heinz 2011). With product heating being a big user of energy, the company has found ways to conserve energy in this key area. Heat recovery systems have been a valuable approach. A ketchup process uses recovered heat to warm water to reduce steam requirements. Important wins for the company were also achieved with the reengineering of the production lines that heat water for bean hydration and steam generation for bean cooking in their cans (Heinz 2011). Before the energy efficiency effort in the canned bean plant, sterilization of the packed and sealed cans consumed a third of the factory's energy (Carbon Trust 2010). The process was modified to reduce this energy demand. One change was to reengineer water mixing in the hydrostatic sterilizers so cooler water did not mix with heated water which reduced the need to reheat the water—halving the steam consumption in the sterilizer (Carbon Trust 2010). This approach was adopted in other parts of the plant to reduce the need to heat other water streams.

Heinz has also integrated on-site renewable energy generation into their program. The company aims to increase renewable energy utilization by 15% by 2015 (Heinz 2011). Projects have used biomass boilers fueled with food by-products (corn straw, cane bagasse), biogas from wastewater digesters to fuel steam production, and rice husks to help fuel boilers (Heinz 2011).

The company is actively progressing toward important goals to conserve energy and use more renewable energy, both of which help reduce GHG emissions for the company.

## 4.2 Water

Food processing constitutes 25% of all water consumption worldwide (Okos). Water is used for a variety of purposes in manufacturing including washing, transporting, cleaning, and adding to the product. Sugar cane, for example, needs 10 cubic meters of water to wash one cubic meter of cane even before it gets processed to make sugar (Santucci 2006). Food processing's water demand strains limited freshwater supplies and can exacerbate this by polluting freshwater. The price for water is not typically as significant to companies as energy, but wastewater fees can sometimes be noticeable. The risk of water shortages is also a very real concern for many processors. SABMiller assessed the water risk of some of its brewing facilities to determine which needed attention to address potential issues and sites in California, Texas, and Colorado rose to the top (MillerCoors).

It may be surprising that water is a scarce resource. While water covers approximately 70% of the "blue planet," less than 1% of it is freshwater available for human use. According to the EPA, water use in the United States has tripled in the last fifty years and it continues to increase every year (EPA b). Globally, increases in water use outpace population growth by 200% (United Nations Water). The small amount of freshwater available is not equally distributed across the globe with some areas at a significant risk of water shortages. Due to these issues, at least a third of the U.S. states are faced with local, regional, or statewide water shortages even under nondrought conditions (EPA b). To add to water challenges, freshwater pollution is a concern. Mercury contamination, for example, is a common problem in lakes in the United States. Game fish from 49% of the lakes in the United States have such high levels of mercury (a known neurotoxin) that they should not be eaten (EPA 2010). As a result, there are numerous benefits of conserving water including (Baldwin 2012):

- Reducing the energy needed to heat water and thus reducing the corresponding environmental impacts.
- Reducing the demand on water treatment and sewage systems and reducing their corresponding environmental impacts (e.g., energy use).

- Maintaining the health of natural pollution filters such as downstream wetlands.
- Reducing water contamination caused by polluted runoff from overirrigated landscaping.
- Reducing the demand (and cost) for additional freshwater to meet growing demands and their associated environmental impacts.
- Reducing surface water withdrawals that degrade habitats both in streams and on land close to streams and lakes.
- Retaining a freshwater supply for generations to come.

Water use in food processing is dominated by washing and sanitation needs (see Figure 4.4). Fruit and vegetable processing is a heavy user of water in large part because of the need for more washing than most industries (United National Industrial Development Organization [UNIDO]). Half of water used in fruit and vegetable processing is used to clean the product (UNIDO). Conservation efforts should aim to address the leading uses of water including washing and sanitation, cooling and heating, and process water.

Water efficiency efforts include reducing or replacing water use and reusing or restoring water (see Table 4.5). A milk processing plant implemented no and low-cost solutions that saved 13 million gallons of water a year and $400,000 a year (see Table 4.6). This included finding ways to reuse water in the facility. Reused water can be leveraged for initial washing stages of foods, counter-current washing, facility cleaning, producing steam, and noncontact

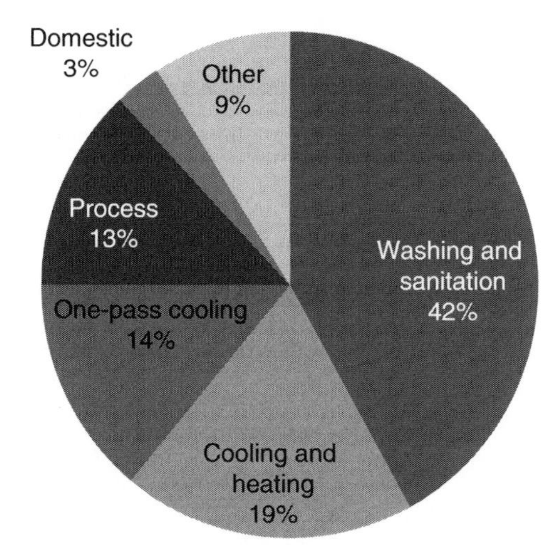


**Figure 4.4** Typical water uses in food processing (reproduced by permission of New Mexico Office of the State Engineer 1999).

**Table 4.5** Water efficiency options (adapted from California Department of Water Resources)

| Minimal Investment ⟶ Larger Investment | | | |
|---|---|---|---|
| Maintenance | Minor Adjustments | Process Adjustments | Efficient Technology & Restoration |
| • Adjust equipment (water pressure)<br>• Fix leaks<br>• Turn off idle equipment<br>• Use dry cleaning before washing | • High-pressure, low-volume nozzles<br>• Springload hoses<br>• Reuse water | • Plan line usage to reduce changeover cleaning | • Upgrade older equipment<br>• Use dry or mechanical conveying and other processes<br>• Restore freshwater (improve wetlands and watersheds, rainwater collection) |

**Table 4.6** Dairy plant case study on water efficiency options that led to 13 million gallons of water saved and $400,000 (University of Minnesota b)

| Minimal Investment ⟶ Larger Investment | | | |
|---|---|---|---|
| Maintenance | Minor Adjustments | Process Adjustments | Efficient Technology |
| • 5,000 gallons a day saved by fixing leaks<br>• 9,400 gallons a day saved by turning off idle water use<br>• 3,000 gallons a day saved by adjusting centrifuge water use<br>• 1,250 gallons a day saved by reducing CIP water | • 14,700 gallons a day saved by reusing cleaning water<br>• 5,300 gallons a day saved with water-efficient showerheads and sprayers | | |

cooling/heating uses (Uyttendaele & Jaxsens 2009). Potential sources of water for reuse include process water (last rinsing water of CIP-cleaning), product water (evaporated), and reconditioned water (treated rainwater) (Uyttendaele & Jaxsens 2009). Leading companies are also involved in restoring freshwater resources through improvement projects to local watersheds and nearby wetlands (see the Coca-Cola example that follows).

After water has been used at a processing plant it is contaminated with residual foods, cleaning and sanitizing chemicals, or other contaminants (e.g., microbes, solids). Biochemical oxygen demand (BOD) and chemical oxygen demand (COD) are used to characterize wastewater quality and can relate to process efficiency (since the materials that end up in the wastewater were not used fully). The BOD, for example, indicates the amount of

**Table 4.7** Wastewater BOD reduction options (adapted from UNIDO)

| | |
|---|---|
| **Source reduction** | • Optimize process efficiency for less food waste<br>• Use dry or mechanical methods instead of wet ones (e.g., scrapers to clean floors and equipment before washing, mechanical conveyance instead of water)<br>• Use high-pressure sprays<br>• Plan line usage to reduce changeover cleaning |
| **Management alternatives** | • Use food waste for animal feed, compost, or land spreading |
| **Advanced technologies** | • Membranes (Microfiltration, UF, RO)<br>• Disinfection (UV, ozone)<br>• Charge separation (ion exchange)<br>• Mechanical separation<br>• Anaerobic digestion<br>• Engineered natural systems or constructed wetlands |

organic material, solids, minerals, nitrogen, and phosphorus in the water from processing. Municipalities have limits on the BOD level before it can be released from a processing facility since high levels can overtax the municipal treatment systems. The BOD in the wastewater can be reduced by optimizing the process for less waste, along with other approaches outlined in Table 4.7.

Anaerobic wastewater treatment may be an alternative as well. This would replace an aerobic wastewater system on-site. The benefits, compared to traditional aerobic wastewater treatment, are highly effective treatment, a lower energy requirement, lower sludge production, and the production of biogas (methane) that can be used as a source of energy (Galitsky et al. 2003).

### 4.2.1 Nestlé

Nestlé is the largest global food company. With such a scale, it is important that the company integrates sustainability into the business. Nestlé's operations are aiming to reduce water use in their facilities and effectively treat the water discharged.

Operational efficiency is a cornerstone of Nestlé's water program. However, to help prioritize the company's efforts water supply risks are assessed. This is done by evaluating water stress based on water withdrawals to availability ratio, estimated annual renewable water supply per person for 2025, and physical risk as determined by the water Risk Filter tool from WWF and the German Development Finance Institution (DEG) (Nestlé a). This helps prioritize water stressed/water scarce areas. These, as well as other sites, have a supplementary local review of the company's potential impact on the communities' right to water and the long-term availability of water. This insight helps the company focus on the areas of greatest need.

Water saving projects in manufacturing helped the company reduce water use by 9% per tonne of product in just one year, between 2011 and 2012, with additional substantial improvements prior to that (e.g., 30% reduction between 1997-2001) (Nestlé a and Nestlé b). The Nestlé Water Policy—Water Resource Guidelines for Sustainable Management is a reference for facilities managers to conserve water (Nestlé c). In addition to supporting the site's compliance to local laws, it helps the sites implement efforts to achieve water use reduction from basic conservation projects to significant advances with new technologies and water reuse systems. A new water treatment system helped reduce water consumption by 30% for a facility in Vietnam (Nestlé a). Twenty-six percent of a Philippine factory's water use was saved by collecting, cleaning, and reusing water for the cooling tower and garden irrigation (Nestlé a).

A factory in Spain implemented the full range of changes (from basic conservation to new technologies and reuse) to yield a dramatic water reduction of 60% in just 12 months (Edie Newsroom 2013). Staff training on behavior changes helped move progress. An adjustment to the vacuum required for a milk evaporator added to it. New cooling towers that recycle water topped off the savings.

Water efficiency, collection, and reuse projects also help address wastewater needs by reducing the total water needed for treatment or discharge. Between 2002 and 2012, Nestlé reduced wastewater by 45% with this approach (Nestlé a). The remaining water is treated to ensure its quality. In addition to technological treatment of wastewater, Nestlé constructed wetlands to manage wastewater at some sites in the United States (Nestlé b).

From water efficiency to wastewater optimization, Nestlé is actively addressing water concerns from their processing operations. Clear achievements have already been shown with approaches that are set up for continued progress.

## 4.2.2 The Coca-Cola Company water stewardship

The well-known beverage company Coca-Cola is dependent on affordable and clean sources of water. Water is a key ingredient for the products and is used for rinsing, heating, cooling, and washing. The company has been working to advance sustainability for the freshwater supply with several approaches. Water use efficiency and wastewater treatment are key manufacturing strategies along with a unique approach to replenish water availability—aimed to reduce water needed to produce the product by 25% by 2020 compared to a 2010 baseline, return process water to a level that supports aquatic life, and safely return to communities and nature an amount of water equal to what is used in the products and production (Coca-Cola Great Britain a).

Coca-Cola addresses their water goals through a better understanding of the opportunities and needs. The company completed a water footprint to identify supply chain issues. While the water footprint highlighted the need to address upstream water issues in the agricultural supply, the company continues to evaluate the different impacts all of their water uses have—especially in geographic areas with water scarcity. This is done by assessing the vulnerability of the quality and quantity of water sources at each bottling plant. A customized plan to address water concerns then gets developed to suit the local needs.

In order to achieve the water reduction goal in manufacturing, Coca-Cola precisely monitors water use with meters in all the factories. Being able to see water use more carefully ended up contributing to substantial improvements in water efficiency. Overall, the company has improved the water efficiency 5.9% since 2010 and 21.4% since 2004 (Coca-Cola 2013). The successes over the years were compiled into a Water Efficiency Toolkit that contains more than 60 best practices to help each site consider potential water conserving opportunities (WWF).

New technology also contributes to Coca-Cola's water reduction efforts. In the UK, water has been replaced with dry options for conveying and even rinsing. Conveyors that move cans and PET bottles along production lines were switched from a soapy water lubricant to a dry lubricant (Coca-Cola Great Britain b). Ionized air is being used to "rinse" preblown bottles and cans instead of water (Coca-Cola Great Britain b).

Water reuse opportunities are being leveraged as well. Many of the company's factories have found ways to circulate water used to wash empty packages or cans to cool pumping equipment (Coca-Cola Great Britain a). A 20,000-liter harvesting system was installed to capture water for reuse for vehicle washing, floor cleaning, and toilets. (Coca-Cola Great Britain b).

Company-developed wastewater standards have been implemented so used water reenters the environment without harming it. In many localities, this is higher than legal requirements. The requirements include a BOD of <50 mg/L, pH 6.5-8, total suspended solids <50 mg/L, total dissolved solids <2,000 mg/L, total nitrogen <5 mg/L, and total phosphorus <2 mg/L, among 14 other parameters (Coca-Cola).

To close the water loop, Coca-Cola replenishes the water used in their products to communities and nature by supporting healthy watersheds and development programs. Over 350 projects to improve access to water and sanitation, watershed protection, water availability, and awareness of water issues helped balance about 35% of the water used in their products corporatewide, from 2005 to 2011 (Coca-Cola).

In India, Coca-Cola harvested rainwater, restored ponds, and assisted water conservation in the community. Water-efficient agriculture production with drip irrigation helped the local farmers reduce water use while improving their yields. The overall effort in India already achieved a neutral balance of the water used by the company and the water replenished.

This comprehensive approach to water management has place Coca-Cola on the forefront of the issue and the company continues to advance their efforts.

## 4.3 Chemicals and other inputs

Cleaning and sanitizing processes in food plants are essential to maintain food safety and product quality. However, these methods often require chemicals that end up polluting the water discharged from the facility. Cleaning chemicals may contain phosphates and other materials that contribute to environmental damage including choking natural aquatic life through eutrophication. Sanitizing chemicals are powerful agents including chlorine, iodine, quaternary amines, peroxide, peroxyacetic acid, and organic acids. These chemicals have different impacts on human health and the environment (see Table 4.8).

**Table 4.8** Food processing sanitizer environmental and health profile (adapted from Pfuntner 2011; Schmidt, and Gaulin et al. 2011)

| Chemical | Efficacy Spectrum | Environmental and Health Profile |
|---|---|---|
| Chlorine-based | • Broad efficacy spectrum<br>• Not effective on biofilms<br>• Efficacy susceptible to various factors (temperature, pH, solids) | • Skin and eye irritants<br>• Respiratory concerns (emerging as a cause of asthma)<br>• Environmental concerns, especially when combined with organic substance from the environment to form trihalomethanes and dioxins |
| Iodophors | • Broad efficacy spectrum<br>• Not effective on spores or biofilms<br>• Efficacy susceptible to various factors (pH, solids, temperature) | • Generally limited issues |
| Quaternary ammonium compounds | • Broad efficacy spectrum, including biofilms<br>• Not effective on spores<br>• Efficacy susceptible to various factors (minerals, pH, solids, cleaning agents) | • Respiratory concerns (emerging as a cause of asthma)<br>• Environmental concerns |
| Peroxyacetic acid (and organic acids) | • Broad efficacy spectrum, including spores and biofilms<br>• Efficacy susceptible to various factors (pH) | • Skin and eye irritants and corrosive<br>• Environmentally preferable since breaks down in water |
| Hydrogen peroxide | • Broad efficacy spectrum, including spores<br>• Not effective on biofilms<br>• Efficacy susceptible to various factors (minerals, pH) | • Skin and eye irritants and corrosive<br>• Environmentally preferable since breaks down in water |

Chlorine-based cleaning chemicals are widely used. However, there is some pressure to minimize residual chlorine in discharge water due to the environmental concerns (UNIDO). Hydrogen peroxide on the other hand is an effective sanitizing agent with a preferable environmental profile. Environmentally preferable cleaning products, including hydrogen peroxide options, have even been certified (e.g., EcoLogo, Green Seal). Thermal treatment with steam or hot water is also an alternative to chemical cleaning and sanitization. Due to the high energy demand for this option it is generally expensive compared to chemicals and should be carefully evaluated for its trade-offs to ensure it provides the intended benefits (Schmidt). Sanitizing and cleaning approaches and products used in processing should meet food safety, quality, and cost requirements while being the most preferable for the environment and workers.

Refrigeration in processing typically uses a chemical refrigerant. Ammonia is the most commonly used refrigerant in food processing plants. It is not considered a GHG like other refrigerants but is a toxic chemical that cannot be used in all systems. Hydrofluorocarbons (HFCs) are alternatives but are very powerful GHGs, with a global warming potential of 124-14800 times more potent than carbon dioxide. Carbon dioxide is an option when in a subcritical or transcritical state (Morawicki 2012). Rapid product cooling is sometimes achieved with direct exposure to carbon dioxide or nitrogen—with nitrogen being preferable since it does not contribute to GHG emissions.

## 4.4 Lean, clean, and green processing

The combination of efficient (lean), safe (clean), and environmentally proactive (green) processing is what the leading companies are practicing including those described in the examples in this chapter. This approach yields important benefits including conservation of resources, risk reduction, and cost savings. Companies are actively working to reduce inefficiencies and addressing safety. The added needs of chemical, refrigerant, energy, and water use are the new areas of emphasis—the green focus areas. The steps to green management in manufacturing generally follow the process outlined for sustainable sourcing of agricultural materials: identify priorities, develop goals, implement program, and monitor progress to continuously improve. The EPA developed an approach that details these steps in the process (see Figure 4.5) (EPA c).

After the organization decides that it is going to address sustainability in its operations, green management begins with assessing the current status and priorities for improvement and developing appropriate targets. Developing a baseline of the current energy and water use and costs, and GHG emissions is critical in this process. The year selected for the baseline is typically the most representative year for which there is accurate data. This considers factors such as acquisitions or divestitures that may limit data availability or what other goals may have been used as a baseline. The baseline is used to understand

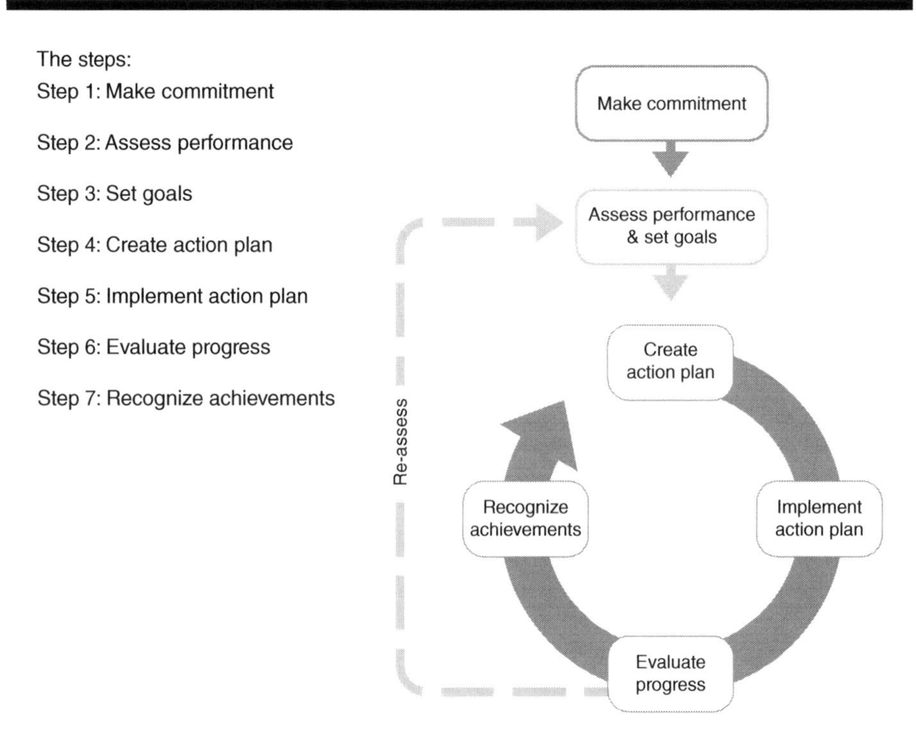


**Figure 4.5** Energy and water management approach (EPA c).

more about the profile of energy and water use and GHG emissions to inform the targets. The baseline is also used to track progress over time. Baseline development is a good opportunity to establish a tracking system for this information moving forward.

Collecting energy and water use data is straightforward, using utility bills and service fees. The GHG emissions baseline requires calculations to determine but there is a well-accepted methodology and factors for this. For processing GHG calculations, scope 1 and scope 2 are typically included (see Figure 4.6) (EPA d). Scope 1 emissions are produced by the business directly; usually including on-site fuel use, refrigerants, and vehicle fuel use. Scope 2 emissions are GHG emissions indirectly produced by the business from its purchased electrical, heat, or steam use. These energy uses are considered "indirect" means of producing GHGs since your business did not generate the energy and its resulting GHG emissions, however, the business required the energy and thus is responsible for the related GHG emissions. Scope 3 emissions are those from waste, employee commuting, travel with non-company-owned vehicles, and emissions from purchases (e.g., emissions from food production, processing, distribution, etc.)—as this is often beyond manufacturing goals it is excluded for this purpose (but may be brought into corporate level goals and assessment). The GHG emissions calculation, often referred to as a carbon footprint, requires data collection for scope 1 and scope 2 items and converting the use to an equivalence of carbon dioxide emissions. These factors are calculated

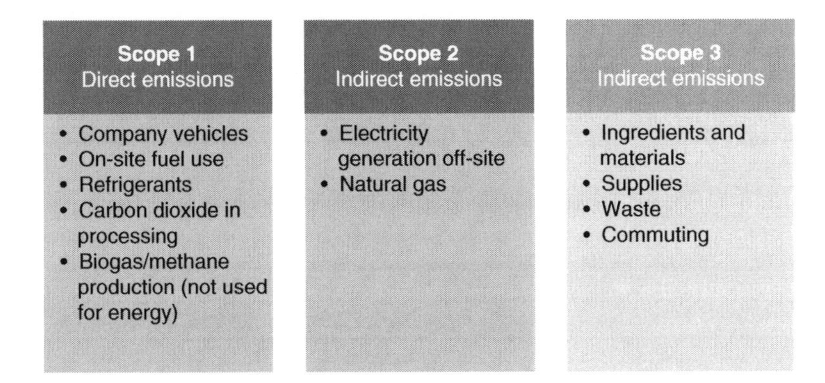


**Figure 4.6** Greenhouse gas emission types/scopes.

ratios relating GHG emissions to a proxy measure of activity at an emissions source (EPA 2005). Actual GHG emissions will include a range of gases (including carbon dioxide, methane, and others). Each gas has its unique effect on the earth's temperature, an effect called global warming potential (GWP). However, it is common practice to express all of the emissions in terms of carbon dioxide as a "reference gas" or base to measure the GWP of each gas. Communicating carbon dioxide equivalent ($CO_2e$) simplifies the discussion of carbon footprints. There are a number of resources and tools that can be used to determine the carbon footprint (e.g., Greenhouse Gas Protocol).

Besides evaluating water use, additional views of water impacts can be evaluated through a water footprint, water scarcity, and water stress. A water footprint is very similar in concept to the carbon footprint. The water footprint is a measure of the water consumed and polluted as a result of your business activities. There are direct uses of water on-site. In addition, the water footprint includes the needed water to produce products and generate electricity for your business, along with other external (indirect) uses of water. Another item included is the amount of pollutants released into the environment. This is because there is a demand for water to dilute the pollutant and bring water quality back to acceptable levels. The technical term used to describe this in a water footprint is "grey water" (Water Footprint Network 2009). Pollutants from food processing can include food waste, grease, cleaning chemicals, pesticides, and other water contaminants. When conducting a complete water footprint, the water needed to "assimilate" these pollutants is counted. Water scarcity and water stress can be assessed using tools such as the World Resources Institute (WRI) Aqueduct tool or the World Business Council for Sustainable Development (WBCSD) GEMI Local Water Tool. These tools leverage available data on locations to identify risks. This insight helps prioritize efforts.

The targets for the program typically are goals for energy, GHG emissions, and water reductions. Most often these range from 3- to 10-year goals with reduction targets of 10 to 15%, however, the goals should be realistic for your organization without being too easy. The EPA has a tool through the Energy Star program to review Energy Performance Indicators (EPIs) that provide an understanding of how other facilities are performing. The United Nations Food and Agriculture Organization estimated that general maintenance in older, less-efficient processing plants can yield energy savings of 10 to 20% with little or no capital investment. With just medium-cost investment savings of 20 to 30% is possible with approaches such as optimizing combustion efficiency, recovering the heat from exhaust gases, and selecting the optimum size of high efficiency electric motors (FAO 2011). Higher savings are possible, but they usually require greater capital investment in new equipment. For more modern facilities the short-term opportunities are fewer, but there are usually still significant gains possible. Benchmarking competitor performance, by looking at their public reporting, may be particularly useful in understanding what is possible. Companies are also utilizing science-based goals that aim for reductions to help meet global targets for GHG emissions or water conservation. The company's capacity for change and resources to support the effort are also important factors in determining appropriate goals. To help with structuring the goals to be achievable, incremental targets can be set that lead to a long-term goal and to encourage and track progress.

The baseline and goals are often normalized by dividing the energy/water use or GHG emissions by the volume or sales or some other relevant factor. This doesn't mean that absolute changes are not tracked and are targeted for reduction, but it is common to also use normalization to account for business fluctuations.

In order to implement an energy, water, and GHG emissions reduction program it is important to develop a plan. The plan should be well-informed by facility audits and available information to identify projects to work on. Table 4.9 outlines key steps in a facility audit and tools to use to collect additional information that is useful to identify facilities to focus on and initiatives to develop. For water audits this might include (EPA e):

1. Identify all water-consuming equipment, high-use areas, and meter locations (process/equipment, cooling/heating, sanitary/domestic, kitchens, landscaping, other).
2. Note all water losses, evaporative losses, and water incorporated in product; excessive water pressure; and leaks (low water pressure, dirty water, and a high quantity of water that cannot be accounted for are all signs of a leak).
3. Observe shift clean-ups and process changeovers.
4. Quantify water flow rates and usage.
5. Note the water quality used in each process step.
6. Determine water quality needs for each process, and quality of wastewater discharged.

**Table 4.9** Tools to identify opportunities and initiatives for energy and water efficiency

| Energy and GHGs | Water |
|---|---|
| • Energy use audit on performance of equipment, processes, and systems (actual versus designed versus best available) by staff or external experts<br>• Carbon footprint<br>• EPA Energy Star tools | • Water audits on performance of equipment, processes, and systems (actual versus designed versus best available) by staff or external experts<br>• Water footprint<br>• WBCSD Global Water Tool/GEMI Local Water Tool<br>• WRI Aqueduct tool |

The projects that are taken on include the range of opportunities discussed earlier in the chapter including maintenance, minor and process adjustment, and efficient technology. Better monitoring systems, such as sub-metering, is a low-cost option that can have significant payback just by being better able to evaluate use. A 5% savings is typical with this approach alone (Galitsky et al. 2003). The plan should ensure there is an appropriate organizational framework of accountability, team, and budget support.

Implementation may also include obtaining necessary capital, setting up the process, or training the staff. Any changes in staff operating procedures need to be made clear with a description of what is being done, how it is being done, and by whom. A key part of the implementation also includes monitoring the initiative to ensure that it is being done and that it is achieving the intended purpose. Typically this monitoring is conducted as part of a broader evaluation of the organization's performance in achieving its energy and water conservation goals. This means reviewing the energy and water use continuously and evaluating performance (additional calculations are needed to track GHG emissions).

## 4.5 Summary

Energy and water use in food processing deplete and damage natural resources. However, there are clear ways to address this, and companies are taking action. The payback from this effort is being realized in cost savings and reduced environmental damage. Progress in energy and water conservation has led to leading companies moving past reduction targets. Top companies are aiming to produce their own energy and to reuse water on-site. Additional efforts from companies such as those included in this chapter point to a path that potentially crosses over to an overall public benefit—more freshwater available. This progress exemplifies the principles, practices, and potential toward a more sustainable food system (Table 4.10).

**Table 4.10** Summary of the principles–practices–potential of sustainability for processing

| | |
|---|---|
| **Principle** | *Food and ingredient processing generates resources and requires minimal additional inputs and outputs* |
| **Practices** | • Conserve, reuse, and provide processing energy and water through lean, clean, and green processing:<br>• Maintenance<br>• Minor adjustments and process adjustments<br>• Efficient technology, renewable energy, and water restoration |
| **Potential** | • Generate renewable energy<br>• Replenish water |

# Resources

EPA ENERGY STAR http://www.energystar.gov/

EPA Carbon Footprint Resources http://www.epa.gov/statelocalclimate/local/activities/ghg-inventory.html

Greenhouse Gas Protocol http://www.ghgprotocol.org/

Science Based Targets http://sciencebasedtargets.org

Water Footprint Network http://www.waterfootprint.org/downloads/WaterFootprint Manual2009.pdf.

WBCSD GEMI Local Water Tool http://www.gemi.org/localwatertool/

World Resources Institute Aqueduct http://www.wri.org/our-work/project/aqueduct

University of Minnesota Energy Efficiency Information http://mntap.umn.edu/food/energy.htm

# References


Baldwin, C. (Ed.). (2012). *Greening food and beverage services: A green seal guide to transforming the industry.* Lansing, MI: American Hotel and Lodging Educational Institute.

California Department of Water Resources. *Water use efficiency ideas.* Retrieved from http://www.water.ca.gov/wateruseefficiency/docs/Food.pdf

Canning, P., Charles, A., Huang, S., Polenske, K., & Waters, A. (2010). *Energy use in the U.S. food system.* Retrieved from http://web.mit.edu/dusp/dusp_extension_unsec/reports/polenske_ag_energy.pdf

Carbon Trust. (2010). *Carbon savings in a can: Efficiencies on the production line.* Retrieved from http://www.heinz.com/csr2011/media/environment/Heinz%20Beanz%20Energy%20Reduction.pdf

Coca-Cola. *Water stewardship.* Retrieved from http://www.coca-colacompany.com/sustainabilityreport/world/water-stewardship.html#section-mitigating-riskfor-communities-and-for-our-system

Coca-Cola Great Britain a. *What is Coca-Cola doing to reduce water waste?* Retrieved from http://www.coca-cola.co.uk/faq/environment/reducing-water-wastage-recycling.html

Coca-Cola Great Britain b. *Water stewardship.* Retrieved from http://www.cokecce.co.uk/media/90799/13759-cce-sb-water-final.pdf

Coca-Cola. (2013). Setting a new goal for water efficiency. Retrieved from http://www.coca-colacompany.com/setting-a-new-goal-for-water-efficiency

Edie Newsroom. (2013). *Nestle yields 60% factory water savings by tweaking production process.* Retrieved from http://www.edie.net/news/4/Nestl--yields-60--factory-water-savings-by-tweaking-production-process/%29

Food and Agriculture Organization (FAO) of the United Nations. (2011). *Energy-smart food for people and climate.* Retrieved from http://www.fao.org/docrep/014/i2454e/i2454e00.pdf

Galitsky, C., Worrell, E., & Ruth, M. (2003). *Energy efficiency improvement and cost-saving opportunities for the corn wet milling industry: An ENERGY STAR guide for energy and plant managers.* Retrieved from http://www.energystar.gov/ia/business/industry/LBNL-52307.pdf

Gaulin, C., Le, M-L., Shum, M., & Fong, D. (2011). *Disinfectants and sanitizers for use on food contact surfaces.* Retrieved from http://www.ncceh.ca/sites/default/files/Food_Contact_Surface_Sanitizers_Aug_2011.pdf

Heinz. (2011). *Corporate social responsibly report 2011.* Retrieved from http://www.heinz.com/CSR2011/environment/

Masanet, E., Therkelsen, P., & Worrell, E. (2012). *Energy efficiency improvement and cost-saving opportunities for the baking industry.* Retrieved from http://www.energystar.gov/buildings/sites/default/uploads/tools/Baking_Guide.pdf

MillerCoors. *Water stewardship.* Retrieved from http://www.millercoors.com/GBGR/Environmental-Stewardship/Water-Stewardship.aspx

Morawicki, R. (2012). *Handbook of sustainability for the food sciences.* Ames, IA: Wiley-Blackwell.

Natural Resources Defense Council. *Risky gas drilling threatens health, water supplies.* Retrieved from http://www.nrdc.org/energy/gasdrilling/

Nestlé a. *Water in our operations.* Retrieved from http://www.nestle.com/csv/water/operations

Nestlé b. *Nestlé and water: Sustainability, protection, stewardship.* Retrieved from http://www.nestle.com/asset-library/documents/reports/csv%20reports/water/sustainability_protection_stewardship_english.pdf

Nestlé c. The *Nestlé water management report.* Retrieved from http://www.nestle.com/asset-library/documents/reports/csv%20reports/water/water_management_report_2006_english.pdf

New Mexico Office of the State Engineer. (1999). *A water conservation guide for commercial, institutional and industrial users.* Retrieved from http://www.ose.state.nm.us/water-info/conservation/pdf-manuals/cii-users-guide.pdf

Northwest Food Processors Association (NFPA). (2013). *Hydrogen fuel cells for power generation—Sierra Nevada Brewing Co.* Retrieved from http://www.nwfpa.org/priorities/government-affairs/218-sustainability/micro-case-studies/environmental-sustainability/1326-hydrogen-fuel-cells-for-power-generation-sierra-nevada-brewing-co

Okos, M. *Developing environmentally and economically sustainable food processing systems.* Retrieved from https://engineering.purdue.edu/ABE/Research/research95/okos.sohn.96.whtml

Pacific Gas and Electric Company (PG&E). (2011). *Sierra Nevada brews environmental leadership.* Retrieved from http://www.pge.com/includes/docs/pdfs/mybusiness/energysavingsrebates/incentivesbyindustry/cs_SierraNevada.pdf

Pfuntner, All. (2011). *Sanitizers and disinfectants: The chemicals of prevention.* Retrieved from http://www.foodsafetymagazine.com/magazine-archive1/augustseptember-2011/sanitizers-and-disinfectants-the-chemicals-of-prevention/

Santucci, L. (2006). *Environmental impacts in food production and processing.* Presentation delivered at Subregional Workshop on the Trade and Environment Dimensions in the Food and Food Processing Industries in the Pacific Suva, Fiji, June 7–8, 2006. Retrieved from http://www.unescap.org/esd/environment/cap/meeting/pacific/presentations/Session%204%20-%20Environmental%20impacts.pdf [16 Jan 2014].

Schmidt, R. *Basic elements of equipment cleaning and sanitizing in food processing and handling operations*. Retrieved from http://ucfoodsafety.ucdavis.edu/files/26501.pdf

Schrope, M. *Fracking outpaces science on its impact*. Retrieved from http://environment.yale.edu/envy/stories/fracking-outpaces-science-on-its-impact

Sierra Nevada. *Sustainability*. Retrieved from http://www.sierranevada.com/brewery/about-us/sustainability

Union of Concerned Scientists. *Coal vs. Wind*. Retrieved from http://www.ucsusa.org/clean_energy/coalvswind/c01.html

United National Industrial Development Organization (UNIDO). *Pollution from food processing factories and environmental protection*. Retrieved from http://www.unido.org/fileadmin/import/32129_25PollutionfromFoodProcessing.7.pdf

United Nations Water. *Statistics: Graphs and maps*. Retrieved from http://www.unwater.org/statistics.html

U.S. Energy Information Administration, United States Department of Energy. (2010). *Annual energy review 2009*. Retrieved from http://www.eia.doe.gov/aer/pecss_diagram.html

U.S. Environmental Protection Agency (EPA a). *Combined heat and power: Frequently asked questions*. Retrieved from http://www.epa.gov/chp/documents/faq.pdf

U.S. Environmental Protection Agency (EPA b). *WaterSense*. Retrieved from http://www.epa.gov/WaterSense/index.html

U.S. Environmental Protection Agency (EPA c). *Guidelines for energy management*. Retrieved from http://www.energystar.gov/buildings/sites/default/uploads/tools/Guidelines%20for%20Energy%20Management%206_2013.pdf?b906-06c1

U.S. Environmental Protection Agency (EPA d). *Greenhouse gas emission reductions*. Retrieved from http://www.epa.gov/greeningepa/ghg/

U.S. Environmental Protection Agency (EPA e). *Lean and water toolkit*. Retrieved from http://www.epa.gov/lean/environment/toolkits/water/resources/lean-water-toolkit.pdf

U.S. Environmental Protection Agency. (2005). *Climate leaders greenhouse gas inventory protocol design principles*. Retrieved from http://www.epa.gov/climateleaders/documents/resources/design-principles.pdf

U.S. Environmental Protection Agency. (2007). *Energy trends in selected manufacturing sectors: Opportunities and challenges for environmentally preferable energy outcomes*. Retrieved from http://www.epa.gov/sectors/pdf/energy/ch3-4.pdf

U.S. Environmental Protection Agency. (2008). *Quantifying greenhouse gas emissions from key industrial sectors in the United States*. Retrieved from http://www.epa.gov/sectors/pdf/greenhouse-report.pdf

U.S. Environmental Protection Agency. (2009). *Clean energy: How does electricity affect the environment?* Retrieved from http://www.epa.gov/cleanenergy/energy-and-you/affect/index.html

U.S. Environmental Protection Agency. (2010). *National lakes assessment: A collaborative survey of the nation's lakes*. Retrieved from http://www.epa.gov/owow/LAKES/lakessurvey/pdf/nla_report_low_res.pdf

U.S. Environmental Protection Agency. (2011). *eGRID2012 version 1.0 year 2009 summary tables*. Retrieved from http://www.epa.gov/cleanenergy/documents/egridzips/eGRID2012V1_0_year09_SummaryTables.pdf

University of Minnesota a. *Air compressor energy-saving tips*. Retrieved from http://mntap.umn.edu/greenbusiness/energy/82-CompAir.htm

University of Minnesota b. *Schroeder milk saves $400,000 through product savings and water conservation*. Retrieved from http://www.mntap.umn.edu/food/resources/80-Schroeder.htm

Uyttendaele, M., & Jaxsens, L. (2009). *The significance of water in food production, processing, and preparation*. Retrieved from http://www.foodprotection.org/events/european-symposia/09Berlin/Uyttendaele.pdf

Water Footprint Network. (2009). *Water footprint manual: State of the art 2009*. Retrieved from http://www.waterfootprint.org/downloads/WaterFootprintManual2009.pdf

World Wildlife Fund. *WWF & Coca-Cola's work to conserve freshwater*. Retrieved from http://worldwildlife.org/projects/wwf-coca-cola-s-work-to-conserve-fresh-water

# 5 Packaging

*Principle: Packaging effectively protects food and supports the environment without damage and waste.*

Packaging is a unique element of food and beverage products. It serves the purpose of transporting and protecting the product, but does not provide nourishment. Packaging has additional functions including portioning, consumer storage, communicating legal and marketing information, and others. As a result, packaging has its own life cycle with material, processing, distribution, and end-of-life considerations. Food packaging accounts for almost two-thirds of total packaging waste by volume (Marsh & Bugusu 2007). This has contributed to putting wasteful packaging on the forefront of environmental concerns for consumers. Yet, over 95% of the environmental impact of packaging is from the production of the package, with the remaining 5% from the disposal. This chapter highlights key topics that contribute to the progress toward more sustainable packaging for food and beverages and points to the opportunity for packaging to have an added function as a valuable resource that can feed into the production of new products.

## 5.1 Packaging hotspots

The early phases in the packaging life cycle place the greatest burden on the environment. This includes the extraction or harvesting of the starting materials and processing those substrates into packaging. Distribution and use of packaging have minimal impacts. End-of-life also has relatively low impacts but represents an opportunity to recover the material with recycling to maximize the use of the available material.

### 5.1.1 Materials

Food and beverage packaging is primarily comprised of plastic (35%), paperboard and corrugate (35%), metals (12%), glass (3%), and other or mixtures of

*The 10 Principles of Food Industry Sustainability*, First Edition. Cheryl J. Baldwin.

materials (Sustainable Packaging Coalition [SPC] 2009a). There may be other materials used for tertiary, or shipping packaging, such as wood for pallets (see Table 5.1). Material selection plays a critical role for the function of the package and also is the source of most of the environmental concerns (see Table 5.2). Comparisons of different materials can be challenging since each material has unique functionality and characteristics such as moisture and oxygen barrier

**Table 5.1** Packaging types and stages (Consumer Goods Forum [CGF] 2011)

| | |
|---|---|
| **Primary** | The packaging designed to come into direct contact with the product. |
| **Secondary** | A group of primary packages/units that are packaged together for retailer handling. Can be a means to replenish the store shelves or a way to sell multiple units on the shelf. |
| **Tertiary** | The packaging used to transport the product, not including any road, rail, ship, or air containers. Typically it is an outer case, pallet, and/or crate. |

**Table 5.2** Typical food packaging materials and their functional and environmental considerations

| | General Functional and Environmental Considerations |
|---|---|
| **Glass** | • Strong and durable<br>• High processing and transportation energy demands<br>• Fully recyclable (without quality loss) |
| **Plastic** | • Versatile, flexible, or rigid<br>• Light in weight<br>• Material sourcing fossil fuel depletion and pollution<br>• High-processing energy intensity<br>• Efficient to transport<br>• Recyclable, but comes with performance loss and typically requires virgin material |
| **Steel and aluminum** | • Strong<br>• Material sourcing mining impacts<br>• High-processing energy and water demands<br>• Variable transportation energy demands (aluminum lighter, less energy needed)<br>• Fully recyclable (without quality loss) |
| **Paper, card, and cartonboard** | • Lightweight<br>• Material sourcing deforestation and related impacts<br>• High-processing energy demand, water use and pollution, and harmful chemical requirements<br>• Recyclable, but comes with performance loss and typically requires virgin material |
| **Mixed materials (e.g., laminates)** | • Versatile<br>• Usual light in weight<br>• Generally not recyclable<br>• (burdens from source materials, see above) |

**Table 5.3** Materials used in food packaging

| Renewable | Nonrenewable | Recycled |
|---|---|---|
| Paper, corrugate | Glass (virgin) | Paper, corrugate |
| Bioplastics | Metal (virgin) | Glass |
| | Plastic (virgin) | Metal |
| | Mixed materials | Plastic |

properties. An understanding of the general functional and environmental considerations can inform design decisions but a more detailed comparison of design options is typically required in order to account for weight or other design differences of the materials (such as through life-cycle assessment [LCA]).

Packaging materials and considerations will be reviewed in the discussion that follows based on their origin of renewable, nonrenewable, or recycled materials (see Table 5.3). It is worth noting, however, that packaging materials are sometimes mixed to deliver certain functionality.

#### *5.1.1.1 Renewable materials*

The primary renewable material used in packaging comes from trees for paper, paperboard, and related materials. Harvesting tress has traditionally come with notable environmental damage. Deforestation causes biodiversity loss, greenhouse gas (GHG) emissions, soil erosion, and water-cycle interruption and pollution. This remains a critical concern (Greenpeace). However, there are responsibly produced and harvested tree fiber options, such as those with a certified chain of custody through the Forest Stewardship Council, among others.

Processing wood fiber requires large volumes of water and powerful chemicals (see Figure 5.1 for an overview of the paper production process and related environmental issues). The process begins with a slurry of 99% water and 1% wood fiber (SPC 2009a). The slurry is then typically treated with sodium hydroxide and sodium sulphide and heated to break down the fibers. The pulp is then bleached and protection and strengthening agents are added depending on the uses for the material—corrugated board and some paperboard is not bleached, for example. Chlorine dioxide is the most commonly used chemical for bleaching. There are undesirable by-products from chlorine dioxide bleaching (also referred to as elemental chlorine free bleaching) including the carcinogen dioxin, albeit fewer than there used to be with older processes. This creates problems since wastewater gets contaminated with dioxin and other materials from the process including organic compounds from the fiber which then can enter the environment (e.g., which can cause dead zones in waterways). The process ends with the pulp then rolled and converted to the form of its final use. This may involve coating or metalizing (typically with lamination) before cutting to size.

Newer and growing renewable packaging options include bioplastics and biopolymers. One of the most widely used bioplastics is polylactic acid (PLA).

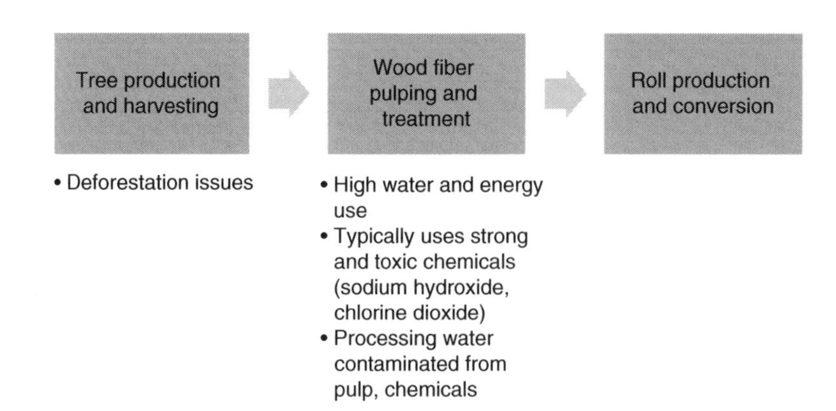


**Figure 5.1** Overview of paper processing environmental considerations.

PLA comprises lactic acid monomers produced by fermentation of corn starch or other sources. It functions similar to fossil fuel-based plastics but is thermally sensitive (softening at less than 105 °F) and has low moisture and gas barrier properties. PLA is used in yogurt, beverage cups, and produce containers. Corn is typically the substrate used in the fermentation and comes with its own environmental burdens (as discussed in a previous chapter) including land and water use and pollution (IFEU 2006; PE Americas 2009). The Bioplastic Feedstock Alliance was formed in response to the growing interest in these materials to guide responsible selection and harvesting of sugar cane, corn, bulrush, and switchgrass used to produce these materials (Lingle 2013a). Other biopolymers are polyhydroxyalkanoates (PHA) produced biologically from microbial polymerization of sugars and sugarcane-derived low-density polyethylene (LDPE). The Coca-Cola Company has invested in the development of another biopolymer, sugarcane-based polyethylene terephthalate (PET) that is recyclable (more details are given later in the chapter).

### *5.1.1.2 Nonrenewable materials*

Nonrenewable materials for packaging are typically mined from the ground or are derived from fossil fuels, both of which are finite resources that are being depleted for packaging production. Glass and metal packaging are derived from mined materials. Glass is made from silica sand, soda ash, and limestone (SPC 2009b). Iron and aluminium are the primary metals in packaging (typically alloyed with manganese, molybdenum, chromium, or nickel for steel) (SPC 2009b). Mining operations have environmental issues from converting land for the mining sites and the associated habitat destruction, water use and pollution, and for metal mining there may also be toxic releases into the environment. Glass mining has fewer environmental issues than metal mining, however.

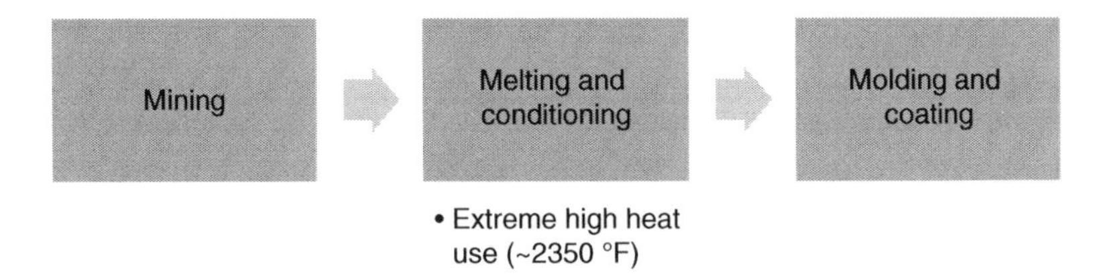


**Figure 5.2** Overview of glass processing environmental considerations.

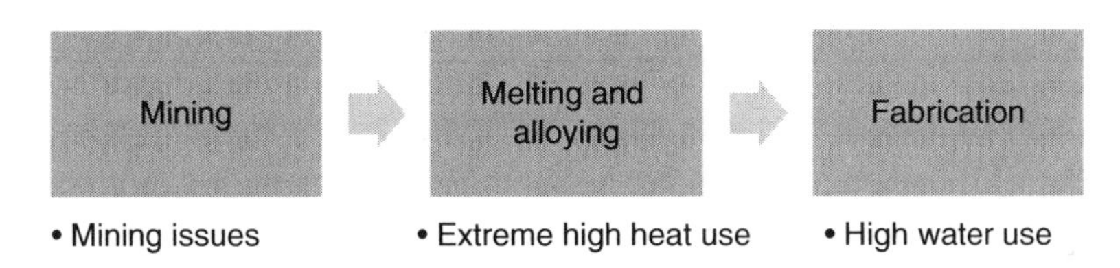


**Figure 5.3** Overview of metal processing environmental considerations.

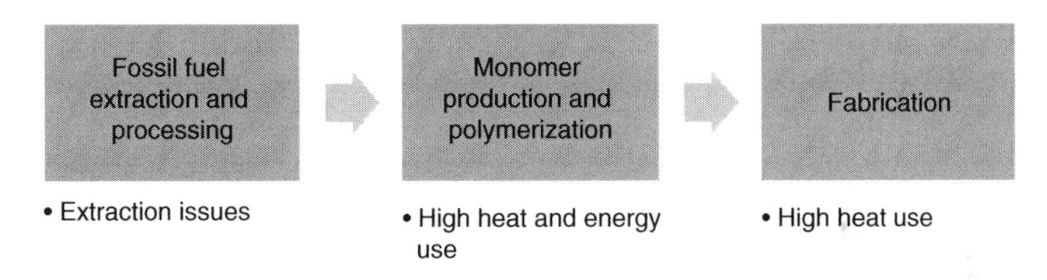


**Figure 5.4** Overview of plastic processing environmental considerations.

Extremely high temperatures are used to produce glass (e.g., 2350 °F) and metal (SPC 2009b). This heat demand dominates much of the environmental impact of these packaging materials (see the general process and associated environmental concerns in Figures 5.2 and 5.3 (The Glass Packaging Institute [GPI] 2010). Glass is considered a bulky and heavy packaging option that comes with transportation impacts from fuel use that are higher than for other materials. However, this only represents a minor contribution to the overall impact of the package, 10% of total energy (GPI 2010). Glass production, with its high heat requirements, dominates the life-cycle burden. Similarly, the heating needed to produce steel (from iron) is so high that it generates 3 to 4% of the total global GHG emissions from all sources (SPC 2009b).

Plastic packaging is made from fossil fuel sources (see Figure 5.4). Spills and leaks on a massive scale are well known (e.g., *Deepwater Horizon* in 2010, *Exxon Valdez* in 1989). These accidents come with visible environmental damage including physical smothering of wildlife with oil and habitat destruction. Less visible are the chemical effects that occur during normal operation from minor leaks and result in contamination with harmful and sometimes lethal effects to both wildlife and humans. (Knoblauch 2009; Natural Resources

Defense Council [NRDC] 2009). In addition, plastic processing also requires high heat to form the material into the package.

Both renewable and nonrenewable materials may have layers of different types of materials added to them for improved functionality. Paper packaging is metallized with aluminum to improve moisture barrier properties. Metallizing paper, or any other material, adds the metal material environmental burden along with the metallizing processing demands (e.g., energy).

#### 5.1.1.3 Recycled materials

Recycled materials are recovered from processing (pre-consumer) or after use (post-consumer). Packaging is an important source of recycled resources. Corrugated boxes used in secondary packaging are the single most recovered packaging material at 72% of generation (United States Environmental Protection Agency [EPA] 2010). Recycled material can be included in new products and packages. Almost all glass packaging contains at least 30 to 35% recycled content, with this expected to increase since up to 70% recycled content is feasible (SPC 2009b). Recovered glass is valuable because it does not lose its quality through the recycling process, unlike paper and plastic, and is not as energy intensive to process as virgin material processing (SPC 2009b). Even at the current low rates of recycled content inclusion, transportation emissions are offset by the energy savings gained with recycled glass (GPI 2010). Similar to glass, steel contains at least 30-35% recycled content and aluminium beverage containers typically have at least 50% recycled content (SPC 2009b). Recycled steel and aluminium also do not lose their quality with recycling and provide an excellent resource for packaging. Steel recycling requires about 60 to 74% of the energy needed to produce virgin steel and aluminum recycling is an even greater savings, close to 99% (EPA; EPA 2005).

Recovered plastic can be used for products and packaging. Pre-consumer and post-consumer plastic is typically washed and ground to flakes for incorporation into the packaging fabrication. Unlike glass and metal that undergo severe heat treatment in reprocessing and do not carry over potential contaminants, recycled plastic has food safety concerns. The U.S. Food and Drug Administration (FDA) reviews each proposed use of recycled plastic because of such issues. The primary concerns with the use of recycled plastic materials for food-contact use include the following (FDA 2006):

1. Contaminants from the post-consumer material may appear in the final food-contact product made from the recycled material.
2. Recycled post-consumer material not regulated for food-contact use may be incorporated into food-contact packaging.
3. That adjuvants in the recycled plastic may not comply with the regulations for food-contact use.

Recovered paper is commonly used in food packaging. Reaching higher levels may be limited by availability and quality of the recycled material. Cereal and pasta boxes contain close to 100% post-consumer content and corrugate typically contains about 40% (both pre- and post-consumer) (SPC 2011a).

The use of recycled material, as with any package, should not lead to adulteration of the food with harmful contaminants from the package. This is especially a concern for recycled paper where there are known contaminant issues (e.g., heavy metals [neurotoxins and carcinogens], BPA, polychlorinated biphenyl [PCB, a carcinogen], toxic mineral oils, and nonylphenol ethyoxylates [NPEs, bioaccumulative and toxic to marine life]) (SPC 2011a). This contamination comes from inks, thermal paper, and carbon paper in the recycling stream as well as the deinking process to recycle paper. Such issues underscore the importance of carefully choosing what goes into the package to begin with (and eventually ends up in recycled content) and selecting the right materials for the right use.

According to the SPC, current guidance from the FDA for recycled content in fiber mirrors the guidelines for recycled plastics content, calling for proof that the paper and paperboard recycling process sufficiently removes potential contaminants (SPC 2011a). The regulations and safety concerns should be carefully reviewed before recycled paper or plastic is used for food packaging and may justify taking a different design approach (e.g., making them more lightweight) (Wiley-Blackwell 2011). If recycled content is used, a close relationship with the supplier that can provide test details and consistent supply is important, but this may not address the concern sufficiently. As a result, Nestlé took this a step further. The company is removing recycled content from food-contact packages (Addy 2013).

### 5.1.1.4 Additives

Inks, glues, and colors are added to packaging to enhance the appearance or structure of the package (Waste and Resources Action Programme [WRAP]). These materials are typically chemical-based and can come with inherent hazards that need to be considered. Inks and adhesives may be highly volatile which can contribute to health and air quality problems. Further, inks and colors may contain toxic heavy metals that need to be limited. Each packaging additive should be carefully reviewed for potential health and environmental issues. Additives may negatively impact recycling processes too. Adhesives are one of the most challenging contaminants to manage during recycling, for example (SPC 2011a). Reducing the overall use of adhesive helps and there are new options such as biopolymer-based adhesives that do not have the recyclability challenges of traditional adhesives (WRAP a).

Plastic requires additives, catalysts, stabilizers, and pigments that can migrate into food (Figge & Freytag 1984). Properties of the plastic influence the potential for migration. The density of the plastic is a key consideration where more migration is likely with the less dense plastics (e.g., LDPE) (Figge & Freytag 1984). Impact modifiers and processing are additional factors that determine the migration potential from the package into food (Figge & Freytag 1984).

Bisphenol A (BPA) has been a common additive to plastics and can coatings. When used in epoxy can coatings, it provides an important barrier between the food and packaging that otherwise might react and contaminate the product. However, BPA also migrates into the food and has undesirable health impacts, disrupting endocrine function leading to reproductive issues and cancer (Biello 2008). BPA has been a target for removal from baby products and packaging for formula and foods. Additional BPA removal is being done, but there are limited alternatives that are known to be safer – companies are being careful not to replace one harmful chemical with another, equally harmful chemical.

### 5.1.2 End of life

Food packaging is a visible form of the food supply's waste. Once the product is consumed, the packaging is left to be handled separately. Consumer packaging waste, however, is not the only source of packaging waste along the supply chain. Packaging waste comes from each stage in the supply chain including restaurants, retailers, distributors, and manufacturers. Not all of the waste is managed appropriately. For example, some of it has ended up clogging our oceans where plastic litter is swirling around choking, trapping, and poisoning wildlife. Used packaging may be recovered for recycling and reuse. This is a critical way to minimize the overall burden from food and beverage packaging and potentially serve as a resource for new packaging.

Paper and paperboard packaging is the largest portion of packaging waste generated in the United States (EPA 2010). Although it is also recovered at the highest rates (greater than 70% recovery), the amount that ends up in landfills contributes to climate change since it releases the potent GHG, methane, when it degrades (EPA 2010). Its rate of decomposition is about 30% versus 1% in 100 years for plastic (Doka 2003). This slow rate of degradation of plastic in the landfill does not warrant the low rates of recovery (at about 13.5% in the United States), especially because recycled plastic can be used for a number of purposes (EPA 2010). Higher recovery rates are possible as evidenced with glass having a recovery rate of 33%, aluminium 50%, and steel 67% (EPA 2010).

Municipal solid waste includes landfills and incineration. In the United States about 12% of waste is incinerated (EPA 2010). Plastic packaging can release toxins when incinerated, requiring control measures to manage (SPC 2009c). PVC releases dioxin and halogens, among other toxins, when incinerated. Other plastics can include heavy metals that contaminate incineration ash to make it potentially hazardous (Marsh & Bugusu 2007; SPC 2009c). Therefore, it is important to consider end of life impacts at the design phase and in program development to reduce issues from arising downstream.

**Table 5.4** Summary of key social considerations for packaging

| Labor | Community |
|---|---|
| • Child labor<br>• Forced or compulsory labor<br>• Freedom of associations and/or collective bargaining<br>• Discrimination<br>• Excessive working hours<br>• Remuneration<br>• Occupational health<br>• Safety performance<br>• Responsible workplace practices (e.g., free of harassment, abuse) | • Compliance with laws<br>• Tenure and use rights and responsibilities (clearly defined, documented, and legally established)<br>• Indigenous people's rights (to lands and resources)<br>• Community relations and workers' rights (economic well-being)<br>• Benefits from the resource (efficient use to ensure economic viability and a wide range of benefits) |

### 5.1.3 Social hotspots

Packaging sourcing and production has the potential to compromise labor standards and community development. These issues tend to be similar across different packaging materials but are of greater concern in developing regions. Table 5.4 outlines the main social considerations for packaging. Working directly with suppliers and selecting certified sources are the best way to address these issues.

## 5.2 Responsible packaging

The environmental and social hotspots with packaging can be addressed to help it be effective, efficient, cyclic, and safe—responsible packaging (or more sustainable packaging) is often defined to include the following (SPA 2010; SPC 2011b):

- Being beneficial, safe, and healthy for individuals and communities throughout its life cycle.
- Meeting market criteria for performance and cost; being sourced, manufactured, transported, and recycled using renewable energy.
- Optimizing the use of renewable or recycled source materials.
- Manufactured using clean production technologies and best practices.
- Made from materials healthy in all probable end-of-life scenarios.
- Physically designed to optimize materials and energy.
- Effectively recovered and used in biological and/or industrial cradle-to-cradle cycles.

**Table 5.5** Summary of approaches for more sustainable packaging (in addition to ensuring the package is functionally effective)

| Materials and Sourcing | Processing and Chemicals | Design and Innovation | End of Life |
|---|---|---|---|
| • Recycled content<br>• Renewable materials<br>• Responsible sourcing | • Clean production<br>• Avoid chemicals of concern | • Light-weighting<br>• New shape<br>• Efficient transport<br>• Edible<br>• Shelf-life extension | • Reusable<br>• Recyclable<br>• Compostable |

However, just one of these attributes should not be addressed in isolation or considered universally better than all others since there may be unintended consequences in the other areas. Recycled content can carry significant contaminant potential that needs to be considered, for example. Table 5.5 summarizes the primary design approaches to improving the sustainability of packaging; also, Table 5.6 outlines detailed criteria from the SPC. It is important to evaluate packaging changes across all of these areas to be sure that potential negative impacts are not being overlooked. This is the approach that Walmart put into place in 2008 with its packaging scorecard. This program evaluated packaging on several sustainability considerations including (Baldwin 2009):

- Greenhouse gases/carbon dioxide tons of production
- Material value
- Product–package ratio
- Cube utilization
- Transportation
- Recycled content
- Recovery value
- Renewable energy
- Innovation

Figure 5.5 shows a process to advance sustainability packaging efforts. It is useful to begin by first understanding the scope of the effort: Will it include a subset of the product portfolio? What aspects of sustainable packaging will be considered (social, environmental, package, economic, etc.)? What metrics are important? Guidance can come from standardized options for metrics, such as from the Consumer Goods Forum, Sustainable Packaging Coalition, and the

**Table 5.6** Sustainable packaging coalition indicators (reproduced by permission of GreenBlue and adapted from SPC 2009d)

| Impact Area | Indicators |
|---|---|
| **Material use** | • Total material use<br>• Material use reduction<br>• Material waste<br>• Virgin material use<br>• Renewable material use<br>• PCR material use<br>• Chain of custody<br>• Production yield |
| **Energy use** | • Total, nonrenewable, and renewable life-cycle energy intensity (with and without transport)<br>• Renewable energy proportion<br>• Recovered latent energy |
| **Water use** | • Life-cycle water consumption<br>• Life-cycle water from stressed sources |
| **Material Health** | • Toxicants concentration<br>• Toxicants migration |
| **Clean production and transport** | • Toxic emissions (total, sulphur oxides, nitrogen oxides, particulate matter)<br>• Life-cycle GHG emissions<br>• Air emissions<br>• Water emissions (total, COD, suspended solids, nitrates)<br>• Environmental management system use<br>• Energy audit conducted |
| **Cost and performance** | • Total cost of packaging<br>• Packaging product wastage<br>• Life-cycle embodied energy production<br>• Packaging service value<br>• Selling unit cube efficiency<br>• Transport packaging cube efficiency |
| **Community impact** | • Product safety<br>• Recycling of packaging<br>• Reuse of packaging<br>• Landfilling of packaging<br>• Packaging energy recovery rate<br>• Packaged product shelf life<br>• End-of-life communications<br>• Community investment |
| **Worker impact** | • Child labor<br>• Forced or compulsory labor<br>• Freedom of associations and/or collective bargaining<br>• Discrimination<br>• Excessive working hours<br>• Remuneration<br>• Occupational health<br>• Safety performance<br>• Responsible workplace practices |

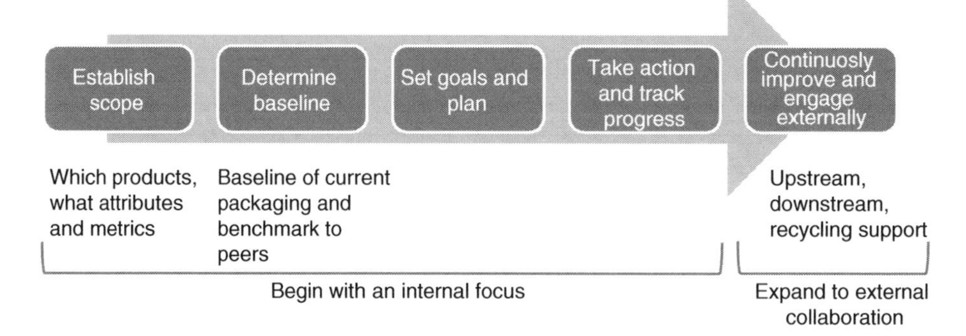


**Figure 5.5** General process to advance sustainability packaging efforts.

Sustainable Packaging Alliance. Any customer request also provides important guidance on metrics.

The next step is to better understand the current situation by developing a baseline with the selected metrics from the scope. Review your baseline with what competitors are doing: consider their current packaging and look at their programs (what are their goals, what pilots do they have). With this information, goals and a plan can be set. Many food manufacturers have goals on improving packaging efficiency and reducing material use. The Coca-Cola Company has a goal to improve packaging material efficiency per liter of product sold by 7% by 2015 compared with a 2008 baseline (Coca-Cola Company 2013). The company has an additional goal to source 25% of PET plastic from recycled or renewable materials by 2015 (Coca-Cola Company 2013). The targets then provide clear direction to take action and track progress. With internal efforts underway, companies then begin to engage externally to improve upstream or downstream issues, including supporting greater material recovery. The Coca-Cola Company has a target to recover 50% of the equivalent bottles and cans used annually by 2015 (Coca-Cola Company 2013).

## 5.2.1 Materials and sourcing

Material selection is a critical decision that impacts other choices and options. With much of the environmental burden from packaging in the material production stages, this decision often drives the sustainability profile of the package (European Organization for Packaging and the Environment [EUROPEN] 2011). The material choice comes down to balancing renewable, recycled content, and recyclable/compostable options, along with the other design considerations (from Table 5.5).

Including recycled materials in packaging reduces the requirements for new materials and their associated impacts (SPC 2006). Recycled paper pulping requires less toxic chemicals and about 40% less energy compared to virgin pulping, for example (Conservatree; U.S. Energy Information Association [EIA]). There are some technical limitations with using recycled content including important food safety concerns as well as the need for regulatory approval and material availability. It is critical to review the quality of the recycled content to ensure that it is appropriate for the intended use, since as discussed earlier there can be contaminants of concern that need to be avoided (SPC 2011a).

Cereal manufacturer Kellogg launched a new product in 2013 in a 15% post-consumer HDPE pouch (Cuneo 2013). This unique package uses less material than the traditional bag-in-box format and also serves as a resealable bag to reuse when the product is gone (as a freezer bag or for dry goods). The recycled HDPE is FDA-approved and meets a set of purity standards through a patented cleaning process that eliminates contaminants without degrading the resin (Knights 2014). Incorporation of recycled HDPE into the cereal package provides a reduced environmental burden with the recycled resin production responsible for 78% less greenhouse gas emissions and 90% less energy than virgin resin (Envision Plastics).

When recycled content cannot be maximized, biobased materials such as paper are often prioritized over fossil fuel-based materials (SPA 2010). This is in part because they are renewable, recyclable, and inherently have a lower energy intensity than fossil fuel-based materials. However, some bioplastics have presented an end-of-life challenge since they are generally not composted (as desired) and can end up contaminating the plastic recycling stream. An exception is the plant-based PET that is recyclable, used in Coca-Cola bottles (see the details that follow). It is important to consider how the material will be collected at the end-of-life to be sure the best material option is selected.

In 2010, Coca-Cola Company began including renewable material across its beverage packages. The PlantBottle is made from about 30% renewable material derived from sugarcane. The mono-ethylene glycol component of PET comes from this biobased material (Coca-Cola Company 2012). The rest of the PET, purified terephthalic acid, is from conventional fossil fuel sources. The result is a plastic that resembles traditional PET and thus can be recycled along with it. This makes the PlantBottle a renewable plastic with a desirable end-of-life option. The company aims to have the plastic contain more biobased content and to have a 100% renewable and recyclable product.

For any type of material used the social risks from packaging including labor and community issues can be addressed through responsible sourcing programs. Companies often approach this need by assessing the potential risks and prioritizing certain materials or areas to focus on. For example, virgin forest products are particularly vulnerable. Using certification programs such as the Forest Stewardship Council (FSC) can help address labor and community issues. If such a certification program is not available for the material sourced, it is important to review the suppliers' practices to ensure labor and social issues are managed. The SPC recommends considering labor and wage issues, eco-justice and pollution of communities, unfair or unbalanced sighting of hazards, and risks associated with social standing (SPC 2006). To do this, many companies conduct social audits of their suppliers.

One of the largest manufacturers of juice drinks in the United States, Johanna Foods, was the first company to offer FSC-labeled beverage cartons (Mohan 2011). The cartons typically contain 75% pulp from wood. The packages contain FSC-certified pulp to provide assurance that the renewable material is grown and harvested responsibly. The packages are also recyclable and lightweight, providing a preferable packaging option.

#### 5.2.1.1 Processing and chemicals

Clean production of packaging with reduced energy and water inputs and waste avoidance can help address environmental concerns (details on processing are presented in another chapter). This can be further enhanced with the use of renewable energy in production and transportation. Food packaging manufacturer Tetra Pak has a program to conserve energy and increase the proportion of renewable energy for production (Tetra Pak 2009). A Tetra Pak facility was built in China in 2009 with the intent to use 100% renewable electricity (Tetra Pak 2009). This joined several facilities that were already using clean energy alternatives from solar panels to wind (Tetra Pak 2009).

Processes that rely on harmful chemicals, such as paper bleaching, should be minimized. Companies should also avoid specific chemicals of concern that are added to packaging (e.g., phthalates, BPA). A more proactive approach would be to conduct a careful assessment of all packaging additives to ensure health and environmental concerns are not overlooked. Evaluating migration of these materials into the food and addressing any potential issues is essential even for packaging that is not in direct contact with the food since migration can occur through layers of packaging or otherwise contaminate food (Everts 2009; Munckel 2013). For example, the printing ink chemical, isopropylthioxanthone, from the outside of the package ended up in baby formula in 2005 prompting the replacement of the ink (Everts 2009). The challenge often is that the composition of the materials may not be known. As a result, it is important for companies to have a better understanding of the materials in their supply chain (ideally down to 100 ppm) to ensure that they are the safest possible (SPC 2006). The SPC suggests carefully reviewing the following additives (SPC 2006):

- Plasticizers (to improve flexibility)
- Heat stabilizers
- Compatibilizers (to allow two or more materials to coexist in a polymer)

- Dyes and pigments
- Fillers (to improve strength)
- Light/UV stabilizers
- Antioxidants
- Flame retardants
- Inks
- Coatings
- Adhesives

When leading U.S. yogurt manufacturer Stonyfield Farms developed a new PLA container, the company aimed to ensure that undesirable chemicals did not end up in the product as a result of leaching from the package (Pure Strategies). The company's suppliers did not want to share additive and material composition, however. With the help of the consultant group Pure Strategies, Stonyfield developed a Safe Additives Guide for its suppliers. This guide includes a list of banned additives owing to health and safety concerns. Stonyfield suppliers must verify that none of the chemicals in the guide are present in their materials. This gives Stonyfield the assurance that their consumers are getting the safest product.

### 5.2.2 Design and innovation

Package size, shape, and function are key considerations for design. The design of the package drives the ultimate amount of material needed and related impacts. EUROPEN noted that packaging designs that increase shelf life but require more use of material are only justifiable if they prevent product waste (EUROPEN 2011). Reducing the weight of the packaging materials used is a clear path to address environmental issues. Less material means a lower demand for extraction, processing, distribution, and end-of-life handling. However, making materials more lightweight cannot be done in a way that compromises the integrity of the package or increases product waste. Light-weighting has been successful in the past, reducing the weight of milk containers 90% by moving from glass to plastic and beverage cans 35% with material reductions (Baldwin 2009). Reducing packaging weight and packaging-to-product weight ratio are common target indicators for sustainable packaging design. Eliminating unnecessary material or layers or down-gauging materials are ways to optimize material efficiency (SPA 2010).

Removing void space and increasing product concentration, if possible, also effectively improve packaging material efficiency. In some cases the choice of container size may have more environmental impact than either the choice of material or the manufacturing process. In a study conducted by the Center for Sustainable Systems at the University of Michigan, it was found that 32-ounce yogurt containers consumed 27% less energy to produce and distribute than 8-ounce containers (Brachfeld, Dritz, Kodama, Phipps, Steiner, & Keoleian 2001).

Improving distribution efficiencies (e.g., cube utilization) touches on many of the sustainable packaging strategies discussed. Reducing weight helps save fuel and reducing size can maximize the amount of product that can fit in a shipment. To effectively address transportation considerations secondary and tertiary packaging needs to be reviewed. There may be design opportunities similar to primary packaging, especially for reuse or reduction.

Global food manufacturer Kraft Foods developed a lightweight and innovative packaging approach for its commercial salad dressing in 2012. This package shifted from a standard rigid HDPE gallon jug to a flexible film container. The change resulted in 60% less plastic, 70% less greenhouse gas emissions, and 50% less energy (Vijayaraghavan 2012). Food service users of the product were also more satisfied since the package was easier to handle and dispense. More product was able to be extracted from the package to lead to 3% less wasted salad dressing (Pettit & Fredholm Murphy 2012). The company used life-cycle assessment to evaluate their innovative design to ensure that it did not have undesirable trade-offs. There were many benefits with the redesign from more efficient transportation and use of material to less product waste. This packaging win provided the company motivation to incorporate more sustainability considerations into its development efforts (Phansey 2012).

#### 5.2.2.1 Edible packaging and shelf-life extension

Innovations that give packaging new functions are being explored. This includes edible packaging and packaging that more actively interacts with the contents to extend product shelf life.

Edible packaging is typically a thin layer or film that is biodegradable (Marsh & Bugusu 2007). These films are made from corn zein, whey, collagen, and gelatin (Marsh & Bugusu 2007). The key disadvantages of edible packaging are cost and limited functionality. In addition, most edible packaging is hydroscopic. Edible packaging typically does not replace all conventional functional packaging needs. As a result, these have not found wide application.

Including technology to further extend product shelf life is an additional area of innovation. This goes beyond the traditional approach of adding controlled or modified atmosphere within the package to reduce respiration/degradation of the food by flushing the package with specific gases (e.g., nitrogen, carbon dioxide, and oxygen). The next generation is packaging that interacts with its contents. PET bottles that are treated with oxygen scavengers such as iron can extend the shelf life of oxygen-sensitive products including beer or fruit juice (McTigue Pierce 2014a). Besides these packages costing more, the other design considerations (e.g., chemicals, end of life) need further evaluation to determine if they offer a sustainability advantage.

## 5.2.3 End of life

The traditional waste hierarchy of reduce-reuse-recycle comes into play with designing to address end-of-life opportunities.

### *5.2.3.1 Reusable packaging*

Reusable packaging is widespread within the supply chain for pallets, trays, and bulk shipping. These packages are circulated for multiple uses, extending the life of the package. The success of these resource-saving options has been partly due to the ability to economically recover them within the supply chain with existing infrastructure (e.g., delivery). The cost of these reusable packaging options typically provides a savings compared to disposable options over the life of their reuse. Mondelez International, manufacturer of snacks, cookies, and coffee, switched to a reusable rigid package within its factories for storing various products moving between production stages (WRAP 2013). The shift to the reusable package allowed for the use of an automated handling system. The overall initiative saves 1,000 tonnes of cardboard a year, 40,000 road miles a year, and 75 tonnes of carbon dioxide per year (WRAP 2013). The reusable packaging has an operating life of at least five years and is recyclable at the end of life.

However, primary packaging reuse has been limited by the lack of economical returns. When financially feasible systems are in place, such as for local companies including dairies or brewers, the environmental savings can be significant. Regardless of if the package is primary, secondary, or tertiary, effective sanitization is critical to avoid microbial contamination (Embree 2013).

Another approach with reusable packaging is to provide a means for consumers to fill their original package with more product. The refill product comes in a significantly down-sized package, for an overall packaging reduction when the original and refill packaging is combined. Mondelez International launched a refill pouch for their Kenco coffee line, along with a lighter weight design for the standard package (WRAP 2013). The refill pouch reduced weight by 97% and the lighter weight required 7% less glass (WRAP 2013).

### *5.2.3.2 Recyclable packaging*

The ability to recover packaging materials is an important opportunity to improve the sustainability profile. As discussed earlier, there can be significant environmental savings from using recycled content. There is also strong consumer interest in more recyclable packaging (Shelton Group 2012).

The SPC states that recycling is an essential strategy for sustainable cycling of packaging materials, especially nonrenewable materials (SPC 2006). Further, the availability of quality, recycled materials is one of the biggest challenges with using more of these materials in products and packages. WRAP notes that 20 to 30% recycled PET is being used for food packaging when 50% is possible if quality recycled content were available (Dvorak, Kosier, & Fletcher 2013). This strategy includes supporting greater recovery of quality material and designing the package for recycling.

Companies have a critical role in this by supporting the development of recycling infrastructure and by helping consumers understand what packaging is recyclable. There is a gap in the amount of material that can be recycled and what is collected, especially in the United States. In 2008, 44% of packaging was recycled in the United States while Germany recycled 71% and the United Kingdom 62% (SPC 2011c). Further, sorting capabilities are typically insufficient (with too much contamination in recycled material). However, the burden often falls on governments to fund this development. Companies can facilitate resource recovery and quality recycled content availability by funding some of the infrastructure and supporting efforts to expand capabilities (see the PepsiCo discussion at the end of this chapter). Where there is collection consumers need to participate more fully. A simple way to contribute to this is to provide consumers with clear guidance on what they can recycle. Programs in the United Kingdom (WRAP) and the United States (SPC) have developed standardized recycling and waste management labeling that companies can use on packages (SPC; WRAP b).

Wegmans was the first food retailer in the United States to begin using the "How2Recycle Label" in 2014. The company began using the label on their plastic carryout bags that are made with 40% recycled content (McTigue Pierce 2014b). The company has collected used bags in its stores since 1994 and added the new recycling label to further facilitate recycling (McTigue Pierce 2014b). The collected bags are used to make new bags for the company (Wegmans 2014b).

Designing the package for recyclability is another important step. It begins with selecting commonly recyclable materials. However, this does not necessarily mean that the package will be recyclable or should be recycled. It is important that incompatible materials are not mixed in a package. WRAP outlined the following principles of designing for recyclability (WRAP 2009):

- Use fewer packaging materials, in any one pack, to allow for ease of recycling.
- Use single materials where possible.
- Design for ease of separation to allow recycling.
- Clearly communicate the pack material and recycling message.

- Design packs in such a way as all the contents can be easily extracted; reduces waste and eases recycling.
- Avoid use of logos and icons that could confuse the consumer.
- Avoid use of extraneous or additional items on primary packs as they may contaminate recycling streams

Additional details on designing specific materials or package types for recyclability are available from the following organizations:

- WRAP
- SPC
- Association of Postconsumer Plastic Recyclers' Design for Recyclability™ Guidelines for Plastic Bottle Recycling

Plastic pouches used for frozen food have replaced bag-in-box formats to reduce material and weight. These pouches, however, typically contain multiple materials in order to withstand the cold temperatures and to provide high-tear strength, puncture resistance, and barrier properties (Lingle 2013b). Tyson Foods in partnership with Dow Packaging and Specialty Plastics developed a single material pouch made from polyethylene (PE) that allows the package to be recyclable.

#### *5.2.3.3 Compostable packaging*

An alternate means of recycling resources in packaging is composting. Some renewable materials such as paper and PLA can be composted through the decomposition of the material with microbial processes. There are industrial and home-scale composting methods. However, composting waste is not a common practice, especially at the consumer end of the supply chain. Compostable biopolymers can be confused for traditional plastic that is recyclable. As a result, biopolymers contaminate recycling streams. Clear labeling is important to avoid this. Due to these difficulties, efforts to improve the other design options may be a higher priority, especially for consumer products. There are some situations where composting may be a preferable end-of-life option, such as for packages that have no other desirable path (reuse or recycling). This, however, generally is not considered the best use of most resources (EUROPEN 2011).

PepsiCo launched the first compostable chip bag in 2010. The multilayer bag was made from 90% PLA and it could degrade in a hot and active compost pile in about 14 weeks (Frito-Lay 2009). PLA provided the product protection and offered an alternate end-of-life option for packaging that is normally disposed of in the landfill. The package was certified compostable by the Biodegradable Products Institute. Consumers did not readily accept the new packaging since it was noticeably louder when handled. The company modified the adhesive used to bind the layers of the bag to help reduce the noise from the more rigid PLA material (Skidmore 2011). The adhesive muffled the noise to make it more in line with traditional bags.

#### *5.2.3.4 California rigid plastic packaging program*

California enacted the Rigid Plastic Packaging Container law in 1991 to reduce the amount of plastic waste disposed in the state's landfills and to increase the use of recycled post-consumer plastic (CalRecycle 2013a). This directly impacts the food industry. Food products that are packaged entirely in plastic with a relatively inflexible shape or form, a capacity of 8 ounces to 5 gallons, and at least one closure have to be one of the following (CalRecycle 2013b):

- Made with 25% post-consumer material
- Source-reduced (e.g., 10% less weight for container, 10% more concentrated product, or a combination)
- Routinely reused at least five times where the reuse is to hold a replacement product
- Routinely returned to and refilled by the product manufacturer or its agent at least five times to replenish the contents of the original package
- Recycled at a 45% recycling rate
- Part of a company program that uses post-consumer material generated in California that is equivalent to or exceeds 25% post-consumer material.

### 5.2.4 PepsiCo's sustainable packaging program

PepsiCo is a global packaged food and beverage company with a comprehensive commitment to sustainability. The company's aim for packaging is to make it increasingly sustainable, minimizing the impact on the environment. This includes a focus to reduce, recycle, use renewable sources, remove environmentally sensitive materials, and promote the reuse of packaging.

A priority for the company is to achieve "best-in-class" material reduction (PepsiCo a). Over the five-year time period leading to 2012, the company removed 350 million pounds of packaging from products (PepsiCo b). This exceeded the goal by more than 20%. Half the plastic from the 16.9-ounce bottle for Aquafina water was removed since 2002 (PepsiCo b). Other products such as Gatorade, Lipton, and flavored Aquafina also had significant resin reductions. Light-weighting has some limits for flavored or carbonated beverages that currently rely on a thick plastic wall to protect ingredients from exposure to oxygen and to retain the carbon dioxide for carbonation, however, PepsiCo was able to employ new technology for thinner walls and found additional savings with a label that is 10% smaller (McKay 2008).

Another key part of the program is increasing recovery of beverage packaging. The company has made great strides in this effort. Since 2010, PepsiCo added more than 5,000 recycling systems in North America, reaching more than 42 states (PepsiCo b). These new systems recycled more than 196 million beverage containers since 2010 (PepsiCo b).

PepsiCo aims to utilize more of the recovered plastic in its own packages. The company uses 10% recycled material in its plastic soda bottles and is working on increasing that amount (McKay 2008).

PepsiCo also helps consumers divert snack bags from the landfill through TerraCycle brigades. TerraCycle repurposes hard-to-recycle packages into innovative products such as shopping bags. PepsiCo-sponsored collections have diverted close to 30 million packages from the landfill (PepsiCo b).

PepsiCo is recognized for its innovative approach to improving the sustainability profile of packaging. A top example is the SunChips snack bag made from compostable PLA that was discussed earlier in this chapter. Biopolymers have been piloted in beverage containers, including the first 100% biopolymer bottle (PepsiCo c). The company has also pushed recycled content with Naked beverages in 100% recycled PET bottles that are squared off to add efficiencies in distribution (Naked Juice).

PepsiCo addresses responsible sourcing of packaging material. A primary focus is to purchase only responsibly sourced wood fiber. PepsiCo has moved to require suppliers to commit to utilizing one of following forestry standards: CERFLOR, CSA, FSC, PEFC, and SFI (PepsiCo).

As a result, PepsiCo has effectively improved its packaging and continues to take its progress to new levels.

### 5.2.5 Sustainable Packaging Coalition

The collaborative working group in the United States, the Sustainable Packaging Coalition (SPC), helps shape and improve packaging across the industry. Run through the nonprofit organization, GreenBlue, SPC's research provided a critical foundation with the formation of an industry-accepted definition for sustainable packaging. This led to the development of a science-based packaging assessment tool and informative resources for the industry. The organization is also involved in improving material recovery to provide sustainable material options for packaging designers.

Sustainable packaging design involves considering a number of factors at the same time including material selection and sourcing, processing and chemicals, design, and end of life. Life-cycle assessment (LCA) with data-driven models of packaging options can be utilized. However, this approach is often time-consuming and complicated for many users. SPC developed a streamlined LCA tool for packaging, the Comparative Packaging Assessment (COMPASS). The software tool is tailored for rapid packaging design evaluations. Users can compare design options on environmental considerations (fossil fuel use, water, greenhouse gas emissions, etc.), waste, and material attributes (SPC b). This helps balance the many considerations to improve the sustainability profile of packaging. SPC also supports packaging design with a number of resources including metrics to evaluate packaging sustainability, environmental briefs on each material option, and design guides.

To further support sustainable packaging design, the SPC is working to improve the recovery of used packaging with its How2Recycle Label program. This is a voluntary, harmonized on-pack label designed to clearly communicate to consumers how to recycle or otherwise manage used packaging. The label was developed because there are variations in recycling programs across the United States leading to confusion about what is accepted for recycling. Companies have contributed to this situation with unclear package labeling. The How2Recycle Label addresses these issues with a standardized on-pack communication approach. There is also a consumer website that provides additional information. The label was launched in 2012 with twelve companies, including the food and beverage manufacturers General Mills, Kellogg's, and Minute Maid/Coca-Cola. The program has since expanded because the label is well-understood by consumers and leads to the desired consumer action, and has the added benefit of improving the consumer's view of the companies using the label (SPC 2013).

SPC also addresses additional issues across its collaborative working group providing an important mechanism for advancing sustainable packaging.

## 5.3 Summary

Packaging serves a critical role in the food supply to protect and deliver safe food to consumers. Packaging design has progressed to demonstrate the potential to shift the model of packaging being a visible reminder of the waste and inefficiencies in the supply chain to instead be a resource and overall contributor to the supply chain. The best way to get there is to address all of the design approaches including materials and sourcing, processing and chemicals, design and innovation, and end of life. The preferred packaging changes can lead to cost savings to the business such as with less material use or switching to recycled content in some cases where recovered materials cost less than virgin (e.g., resins). Attention to the safety of the food needs to remain a priority, especially with regard to the migration of packaging components. In addition, the recovery of packaging materials demands far greater

**Table 5.7** Summary of the principles–practices–potential of sustainability for packaging

| | |
|---|---|
| **Principle** | *Packaging effectively protects food and supports the environment without damage and waste* |
| **Practices** | • Optimize each package's:<br>  • Functionality<br>  • Materials and sourcing<br>  • Processing and chemicals<br>  • Design and innovation<br>  • End of life<br>• Support recycling |
| **Potential** | • Close the resource loop through recycling<br>• Safe food supply |

emphasis, requiring packaging changes to maximize recycling as well as improvement to recovery systems that the food industry should support. There is a critical opportunity to close the resource loop with used food and beverage packaging serving as a feedstock for new materials and products. Packaging can then serve as an important resource while also delivering safe food (Table 5.7).

# Resources

Consumer Goods Forum (CGF) Global Packaging Project http://globalpackaging.mycgforum.com/
Sustainable Packaging Alliance (SPA) http://www.sustainablepack.org/
Sustainable Packaging Coalition (SPC) http://www.sustainablepackaging.org
WRAP http://www.wrap.org.uk
http://www.wrap.org.uk/sites/files/wrap/Packaging%20Resource%20Summary.pdf

# References


Addy, R. (2013). *Nestle tackles food contact chemicals in recycled paper*. Retrieved from http://www.foodproductiondaily.com/Packaging/Nestle-tackles-food-contact-chemicals-in-recycled-paper

Baldwin, C. (Ed.) (2009). *Sustainability in the food industry*. Ames, IA: Wiley-Blackwell.

Biello, D. (2008). *Plastic (not) fantastic: Food containers leach a potentially harmful chemical*. Retrieved from http://www.scientificamerican.com/article/plastic-not-fantastic-with-bisphenol-a/

Brachfeld, D., Dritz, T., Kodama, S., Phipps, Al, Steiner, El, & Keoleian, G. (2001). Life-cycle assessment of the Stonyfield Farm product delivery system. *Center for Sustainable Systems University of Michigan*. Report No. CSS01-03. Retrieved from http://css.snre.umich.edu/css_doc/CSS01-03.pdf

CalRecycle. (2013a). *California's rigid plastic packaging container (RPPC) program*. Retrieved from http://www.calrecycle.ca.gov/plastics/rppc/#Compliance

CalRecycle. (2013b) *Container compliance options*. Retrieved from http://www.calrecycle.ca.gov/plastics/rppc/Enforcement/Compliance.htm

Coca-Cola Company. (2012). *PlantBottle: Frequently asked questions*. Retrieved from http://www.coca-colacompany.com/stories/plantbottle-frequently-asked-questions

Coca-Cola Company. (2013). *Performance highlights*. Retrieved from http://www.coca-colacompany.com/our-company/performance-highlights

Conservatree. *Environmentally sound paper overview: essential issues part III—making paper: Content*. Retrieved from http://www.conservatree.org/learn/Essential%20Issues/EIPaperContent.shtml

Consumer Good Forum (CGF). (2011). *Global protocol on packaging sustainability 2.0*. Retrieved from http://globalpackaging.mycgforum.com/

Cuneo, L. (2013). Cereal launches in a reusable zippered pouch. Retrieved from http://envisionplastics.files.wordpress.com/2013/07/reusable-zippered-pouch.pdf

Doka, G. (2003). *Life-cycle inventories of waste treatment services*. Retrieved from http://www.doka.ch/13_I_WasteTreatmentGeneral.pdf

Dvorak, R., Kosier, E., & Fletcher, J. (2013). *Improving food grade rPET quality for use in UK packaging*. Retrieved from http://www.wrap.org.uk/sites/files/wrap/rPET%20Quality%20Report.pdf

Embree, K. (2013). *Study: Contamination a concern for containers used to ship produce*. Retrieved from http://www.packagingdigest.com/article/523806-Study_Contamination_a_concern_for_containers_used_to_ship_produce.php

Envision Plastics. *Plastics recycling FAQ*. Retrieved from http://envisionplastics.com/faqs/

European Organization for Packaging and the Environment (EUROPEN). (2011). *EUROPEN green paper 2011*. Retrieved from http://www.europen-packaging.eu/sustainability/packaging-environment.html

Everts, S. (2009). *Chemical leach from packaging*. Retrieved from http://cen.acs.org/articles/87/i35/Chemicals-Leach-Packaging.html

Figge, K., & Freytag, W. (1984). *Additive migration from various plastics with different processing or properties into test fat HB 307*. Retrieved from http://www.mindfully.org/Plastic/Antioxidants/Additive-Migration-Fat1984.htm

Frito-Lay. (2009). *Frito-Lay's SunChips brand changing the future of snack food packaging*. Retrieved from http://www.fritolay.com/about-us/press-release-20090415.html

Glass Packaging Institute (GPI). (2010). *Cradle-to-cradle life-cycle assessment of North American container glass*. Retrieved from http://www.container-recycling.org/assets/pdfs/glass/LCA-GPI2010.pdf

Greenpeace. *How APP is toying with extinction*. Retrieved from http://www.greenpeace.org/usa/en/campaigns/forests/Our-impact/the-breakup/APP-Mattel-and-Rainforests/

IFEU (2006) *Life-cycle assessment of polylactide (PLA): A comparison of food packaging made from NatureWorks PLA and alternative materials*. Retrieved from http://www.ifeu.de/oekobilanzen/pdf/LCA%20zu%20PLA%20erstellt%20fuer%20NatureWorks%20%28Okt%202006%29.pdf

Knights, M. (2014). *Shelf-stable packages win IoPP awards based on composition and design focused on ease of use and sustainable qualities*. Retrieved from http://www.foodengineeringmag.com/articles/print/91716-packaging

Knoblauch, J. (2009). The environmental toll of plastics. *Environmental Health News*. Retrieved from http://www.environmentalhealthnews.org/ehs/news/dangers-of-plastic

Lingle, R. (2013a). *WWF, major brand owners launch Bioplastic Feedstock Alliance*. Retrieved from http://www.packagingdigest.com/article/523909-WWF_major_brand_owners_launch_Bioplastic_Feedstock_Alliance.php

Lingle, R. (2013b). *Tyson Foods debuts the first 100% recyclable stand-up pouch*. Retrieved from http://www.packagingdigest.com/article/523826-Tyson_Foods_debuts_the_first_100_percent_recyclable_stand_up_pouch.php

Marsh, K., & Bugusu, B. (2007). Food packaging: roles, materials, and environmental issues. *Journal of Food Science*, 72(3), R39–R55.

McKay, B. (2008). *Pepsi to cut plastic used in bottles*. Retrieved from http://online.wsj.com/news/articles/SB121004395479169979

McTigue Pierce, L. (2014a). *Food industry focuses on better packaging to cut spoilage*. Retrieved from http://www.packagingdigest.com/article/524024-Food_industry_focuses_on_better_packaging_to_cut_spoilage.php

McTigue Pierce, L. (2014b) *Wegmans is first grocery store to join How2Recycle Label program*. Retrieved from http://www.packagingdigest.com/article/524053-Wegmans_is_first_grocery_store_to_join_How2Recycle_Label_program.php

Mohan, A. M. (2011). *Aseptic juice packs are FSC-certified*. Retrieved from http://www.greenerpackage.com/certifications/aseptic_juice_packs_are_fsc-certified

Morawicki, R. (2012). *Handbook of sustainability for the food sciences*. Ames, IA: Wiley-Blackwell.

Munckel, J. (2013). *Migration*. Retrieved from http://www.foodpackagingforum.org/Food-Packaging-Health/Migration

Naked Juice. *Sustainability*. Retrieved from http://www.nakedjuice.com/our-purpose/#/sustain

Natural Resources Defense Council (NRDC). (2009). *Protecting our ocean and coastal economies: Avoid unnecessary risks from offshore drilling*. Retrieved from http://www.nrdc.org/oceans/offshore/files/offshore.pdf

PE Americas. (2009). *Comparative life-cycle assessment Ingeo biopolymer, PET, and PP drinking cups*. Retrieved from http://www.natureworksllc.com/~/media/The_Ingeo_Journey/EcoProfile_LCA/LCA/PEA_Cup_Lid_LCA_FullReport_ReviewStatement_121209_pdf.pdf

PepsiCo a. *PepsiCo global sustainable packaging policy*. Retrieved from http://www.pepsico.com/Download/Global_Pack_Policy.pdf

PepsiCo b. *Packaging, waste and recycling*. Retrieved from http://www.pepsico.com/Purpose/Environmental-Sustainability/Packaging-and-Waste

PepsiCo c. *PepsiCo develops world's first 100% plant-based, renewably source PET bottle*. Retrieved from http://www.pepsico.com/PressRelease/PepsiCo-Develops-Worlds-First-100-Percent-Plant-Based-Renewably-Sourced-PET-Bott03152011.html

Pettit, D., & Fredholm Murphy, S. (2012). *Foodservice packaging design and sustainable branding through LCA*. Retrieved from http://lcacenter.org/lcaxii/final-presentations/523.pdf

Phansey, A. (2012). *Revolutionizing sustainable product design*. Retrieved from http://www.huffingtonpost.com/asheen-phansey/green-companies_b_1419185.html

Pure Strategies Case Study. PureEnterprise Stonyfield Farm. Retrieved from http://purestrategies.com/case-study/stonyfield-farm

Shelton Group. (2012). *EcoPulse 2012*. Knoxville, TN: Shelton Group.

Skidmore, S. (2011). *SunChips biodegradable bag made quieter for critics*. Retrieved from http://www.huffingtonpost.com/2011/03/01/sunchips-biodegradable-bag_n_829165.html

Sustainable Packaging Alliance (SPA). (2010). *Principles, strategies and KPIs for packaging sustainability*. Retrieved from http://www.sustainablepack.org/Database/files/filestorage/Sustainable%20Packaging%20Definition%20July%202010.pdf

Sustainable Packaging Coalition (SPC a). How2Recycle label. Retrieved from http://www.sustainablepackaging.org/content/?type=5&id=labeling-for-recovery

Sustainable Packaging Coalition (SPC b). *COMPASS: Comparative packaging assessment*. Retrieved from http://www.sustainablepackaging.org/content/?type=5&id=compass-comparative-packaging-assessment

Sustainable Packaging Coalition. (2006). *Design guidelines for sustainable packaging*. Charlottesville, VA.

Sustainable Packaging Coalition. (2009a). *Environmental technical briefs of common packaging materials: Fiber based*. Charlottesville, VA.

Sustainable Packaging Coalition. (2009b). *Environmental technical briefs of common packaging materials: Metals and glass in packaging*. Charlottesville, VA.

Sustainable Packaging Coalition. (2009c). *Environmental technical briefs of common packaging materials: Polymers.* Charlottesville, VA.
Sustainable Packaging Coalition. (2009d) *Sustainable packaging indicators and metrics framework.* Charlottesville, VA.
Sustainable Packaging Coalition. (2011a). *Guidelines for recycled content in paper and paperboard packaging.* Retrieved from http://www.sustainablepackaging.org/Uploads/Resources/recycled-content-paper-packaging.pdf
Sustainable Packaging Coalition. (2011b). *Definition of sustainable packaging.* Retrieved from http://www.sustainablepackaging.org/content/?type=5&id=definition-of-sustainable-packaging
Sustainable Packaging Coalition. (2011c). *Labeling for package recovery.* Charlottesville, VA. Retrieved from https://www.sustainablepackaging.org/resources/default.aspx
Sustainable Packaging Coalition. (2013). *How2Recycle: The How2Recycle Label soft launch report.* Retrieved from http://gb.assets.s3.amazonaws.com/files/how2recycle/How2RecycleSoftLaunch.pdf
Tetra Pak. (2009). *Cutting the impact of food processing and packaging.* Retrieved from http://www.tetrapak.com/documentbank/Climate_Savers_TetraPak_FactSheet_2009.pdf
U.S. Energy Information Administration (EIA). *Using and saving energy.* Retrieved from http://www.eia.gov/kids/energy.cfm?page=3
U.S. Environmental Protection Agency (EPA). *Tools to reduce waste in schools: appendix M.* Retrieved from http://www.epa.gov/osw/education/pdfs/toolkit/tools-m.pdf
U.S. Environmental Protection Agency. (2005). *Waste management and energy savings: Benefits by the numbers.* Retrieved from http://epa.gov/climatechange/wycd/waste/downloads/Energy%20Savings.pdf
U.S. Environmental Protection Agency. (2010). *Municipal solid waste generation, recycling, and disposal in the United States: Facts and figures for 2010.* Retrieved from http://www.epa.gov/waste/nonhaz/municipal/pubs/msw_2010_rev_factsheet.pdf
U.S. Food and Drug Administration (FDA). (2006). *Guidance for industry: Use of recycled plastics in food packaging: chemistry considerations.* Retrieved from http://www.fda.gov/Food/GuidanceRegulation/GuidanceDocumentsRegulatoryInformation/IngredientsAdditivesGRASPackaging/ucm120762.htm
Vijayaraghavan, A. (2012). *Kraft Foods reduced packaging and travel miles.* Retrieved from http://www.triplepundit.com/2012/04/kraft-foods-reduces-packaging-travel-miles/
Waste and Resources Action Programme. (WRAP a). *Material specifics*, Retrieved from http://www.wrap.org.uk/content/material-specifics
Waste and Resources Action Programme. (WRAP b). *On-pack recycling label.* Retrieved from http://www.wrap.org.uk/content/pack-recycling-label
Waste and Resources Action Programme. (2009). *An introduction to packaging and recyclability.* Retrieved from http://www.wrap.org.uk/sites/files/wrap/Packaging%20and%20Recyclability%20Nov%2009%20PRAG.pdf
Waste and Resources Action Programme. (2013). *Courtauld commitment 2: signatory case studies.* Retrieved from http://www.wrap.org.uk/sites/files/wrap/CC2%20case%20studies%20-%20Dec%202013.pdf
Wegmans. (2014). *Sustainability: "Return to sender" bags at Wegmans shine a light on pathway to recycling.* Retrieved from http://www.wegmans.com/webapp/wcs/stores/servlet/PressReleaseDetailView?productId=775604&storeId=10052&catalogId=10002&langId=-1
Wiley-Blackwell. (2011). *Recycled cardboard in food packaging risks contamination.* Retrieved from http://www.disabled-world.com/fitness/nutrition/foodsecurity/recycled.php

# 6 Distribution and Channels

*Principle: Food and ingredients are efficiently delivered across the supply chain and to the consumer.*

Distribution and marketing channels may not be leading sources of sustainability challenges (see Figure 6.1), but they play a critical role in connecting the consumer to the food supply chain. The top food retailer, Walmart, estimated that 90% of its environmental footprint is in the supply chain of its products (Walmart). Life-cycle assessment research on food services illustrated that approximately 95% of impacts studied come from the food purchased (Baldwin, Kapur, & Wilberforce 2010). Previous chapters reviewed the opportunities to reduce the burdens from products and packaging. The global food system has such a massive scale, however, that even lower tier issues are important to address. The transportation of goods and marketing channels can reduce their impact and provide an important contribution to improving the overall supply chain efficiency and effectiveness. There are also opportunities to go further with producing renewable energy and educating the consumer to provide an even greater benefit to the supply.

## 6.1 Transportation

Transporting food by water, rail, road, and even air are all part of our food system. In the United States food travels 1,500 miles, on average, from farm to consumer (Pirog & Benjamin 2003). This is dominated by truck transport (with 71% of total freight transport done with trucks in the United States) (Greene 2011). Food travels more (tonne-km or ton-mile) than any other commodity in the United States—even more than petroleum, chemicals, and wood products (Morawicki 2012). This transport distance comes with

*The 10 Principles of Food Industry Sustainability*, First Edition. Cheryl J. Baldwin.

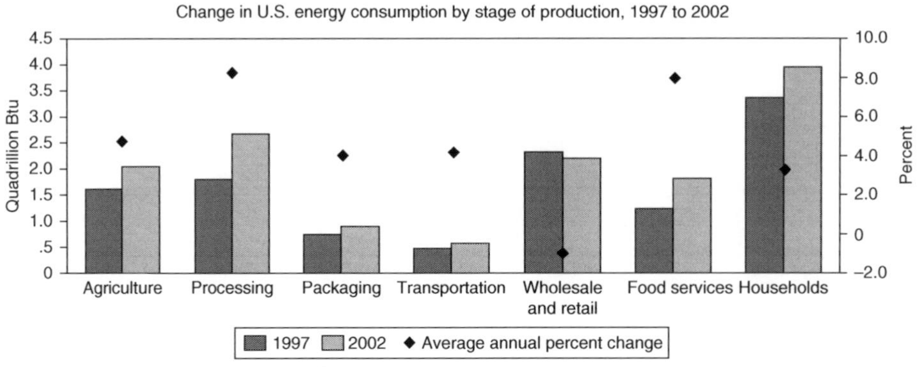


**Figure 6.1** Energy use across the food supply chain (Canning, Charles, Huang, Polenske, & Waters 2010).

**Table 6.1** GHG emission for different transportation modes (Gossling, Garrod, Aall, Hille, & Peeters 2010)

| Mode | GHG (kg $CO_2$/tkm) |
|---|---|
| Boat | 0.015 |
| Train | 0.015 |
| Truck | 0.19 |
| Air | 0.72 |

environmental impacts including climate change, loss of biodiversity, natural resource depletion and degradation, and pollution. Overall, transportation contributes significantly to global greenhouse gas (GHG) emissions with carbon dioxide emitted through the burning of fuels in vehicles. Vehicle fuel also results in air pollutants such as carbon monoxide, hydrocarbons and nitrogen oxides that create smog, sulfur dioxide, and fine particles which all contribute to health problems ranging from headaches, asthma, cancer, to death (U.S. Environmental Protection Agency [EPA] 2007). The construction of roads and fossil fuel extraction/production have their share of issues such as the loss of biodiversity by destruction and disruption of natural habitat. The massive oil leak that lasted for 85 days from the Deepwater Horizon incident in 2010 resulted in 12,000 to 19,000 barrels of oil a day in the Gulf of Mexico (Linkins 2010). The environmental impacts of this accident are not yet fully understood. All of these issues are linked to our use of fossil fuels which are also a limited resource that is rapidly being depleted.

Fuel use drives the environmental burden from food transportation. Much of the fuel impact is determined by the mode of transportation. Table 6.1 shows that air transportation emits the most GHGs, followed by trucks. Boat and train transport have the lowest GHG emissions. Companies are shifting to lower GHG-emitting modes of transport when they can. Ocean Spray in the United States formed a partnership with competitor Tropicana to move product distribution from New Jersey to Florida away from truck and into rail (Kruschwitz 2014). This partnership not only provided Ocean Spray with a way to save costs and GHG emissions but Tropicana also benefited. Tropicana had been sending its rail cars up from Florida to New Jersey full but then they would return back to Florida empty. Now the round trip is fully utilized. In the end, Ocean Spray saved 40% in transportation costs and reduced GHG emissions by 65% and Tropicana eliminated the costs and GHGs from the return trip (Kruschwitz 2014).

Companies looking to reduce their GHG impact from distribution typically pursue a few paths (see Table 6.2):

- Lower GHG-emitting modes (more rail than truck)
- Fuel efficiency measures (maintenance and technology/modifications)
- Route changes (including changing locations of distribution canters to have fewer miles)
- Alternative fuels that emit fewer GHGs

Fuel efficiency can be optimized. Maintenance is a fundamental strategy. For trucks this includes ensuring proper tire pressure, alignment, clean filters, and that there are no leaks (Morawicki 2012). These simple actions can deliver results. Fuel economy, for example, decreases by 1% with every 10 psi of underinflated tires (Morawicki 2012). Driver behavior is another straightforward means of improving fuel efficiency. Improving driving behaviors can reduce vehicle fuel use by 10% and up to 20% for aggressive drivers (United States Department of Energy [DOE a]). Reducing aggressive acceleration, frequent shifting, and idling help. In addition, fuel economy drops with higher speeds. A decrease of 17.3% can be seen with driving at 75 instead of 60 mph (Morawicki 2012).

**Table 6.2** Transportation improvement opportunities

| | |
|---|---|
| **Lower GHG emitting modes** | • Less air and truck<br>• More rail and ship |
| **Fuel-efficiency measures** | • Maintenance (tire pressure, tune-ups)<br>• Driver behavior (speed, acceleration)<br>• Technology/modifications (aerodynamic equipment, idling equipment)<br>• New vehicles |
| **Route changes** | • Route plans<br>• Tracking (GPS)<br>• Location changes (distribution centers)<br>• Inbound logistics optimization |
| **Alternative fuels** | • New vehicles<br>• Conversion of vehicles |

Every 5 mph over 50 mph costs $0.26 more per gallon of gas (based on the price of gas at $3.75 per gallon) (DOE b). Modifications to the truck also help reduce fuel needs with additions to make the vehicle more aerodynamic such as with roof deflectors, cab extenders, chassis fairings, trailer side skirts, underhood air cleaners, concealed exhaust system, aerodynamic mirrors, and trailer end caps (Morawicki 2012). In addition, there are cab heating and electricity options that reduce reliance on idling the engine (DOE c). Direct-fired heaters and auxiliary power units are effective options (Morawicki 2012). In the United States there are truck stops that offer electrified parking spots for the cab to plug into so the truck does not have to idle to run heating and electricity (DOE d). Newer vehicle options provide fuel efficiency advantages with lighter weight, increased volumetric capacity, and more aerodynamic designs (rounded corners and sloped hoods). Similar improvements are available for ships, trains, and planes too.

Paradise Island Foods in Canada implemented a comprehensive driver program that led to idle reduction, slower driving, and better shifting practices (Fraser Basin Council). When it came time to update some of the fleet the company invested in trailers with low emissions engines, speed limiters, on-board driver computers (with GPS and performance tracking), and an aerodynamic package (Fraser Basin Council). Hybrid refrigeration units on the trailers were also included so they can plug into the facility and run on the building's power.

Many companies are actively reviewing routes to optimize trips. This includes reducing miles driven, stops at signals, the number of trucks needed for routes, and time spent in traffic. Fleets typically leverage software applications to optimize their routes. Another advancement is installing GPS-based monitoring systems on trucks to provide real-time data on miles driven, idle time, fuel economy, and engine maintenance information (DOE e). This kind of data helps fleet managers understand not just individual vehicles but also the overall fleet performance to quickly identify opportunities for improvement. With the help of routing and scheduling software, Ireland-based ingredient supplier Glanbia saved 16% in delivery costs (Paragon a). The payback for the software was just six months (Paragon a). Part of the savings came from removal of excess trucks from the fleet and from recalibrating routes and schedules (Paragon a).

There may be opportunities at a companywide level to reduce trips by bringing together shipments based on geography or timing to reduce costs and environmental burden. Review business locations to ensure they are optimized and consider inbound shipments from suppliers, seeking ways to consolidate deliveries. Del Monte Foods opened a distribution center in Topeka, Kansas to create a lower mileage source point for their regional customers (Del Monte Foods). In addition, as production demands increase the company determines the best place for the production (for new and existing facilities) so that finished goods travel the shortest distance to the final destination (Del Monte Foods).

Alternative fuels are a growing path to reducing GHG emissions in truck transport. Alternative fuels include biodiesel, electricity (and hybrid-electric), ethanol, hydrogen, natural gas, and propane (DOE f). These fuels run with fewer GHG emissions than traditional diesel (see Figure 6.2). Ryder Systems' compressed natural gas vehicles, for example, produce up to 27% less greenhouse gas emissions compared to traditional diesel vehicles (Dean Foods 2011). The Coca-Cola Company invested in hybrid-electric trucks with 30% lower GHG emissions for their direct store delivery fleet (Coca-Cola Company). About 10% of the North American fleet are these 35-foot trailer alternatively-fueled vehicles (DOE 2012). Government incentives assisted the acquisition of the fleet. Purchasing new vehicles is one path but vehicle manufacturers also offer aftermarket conversions to enable the switch to an alternative fuel.

Fuel use and GHGs from distribution are common indicators of transportation performance. These are effective since they address multiple variables that need to be considered—mode, fuel, efficiency, and miles. These metrics can be put into terms of the amount of product moved or produced. This is achieved by normalizing (dividing) or by including in a ton-miles per gallon (or tonne-km per liter) metric. Simply looking at food miles is too simplistic. Food miles may have more value when considering local community support (this is discussed in another chapter).

Since as little as 3% of emissions from the food sector likely come from transportation and much of the environmental footprint is a result of production impacts, food transported long distances may have a lower overall

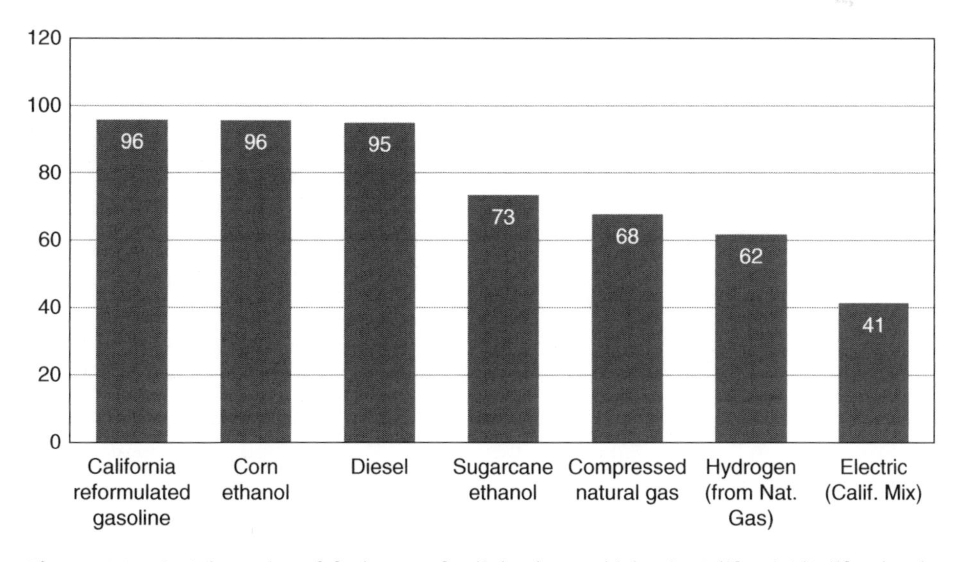


**Figure 6.2** GHG intensity of fuel types for light-duty vehicles in California (California Air Resources Board 2010). *Total Carbon Intensity is given in gCO2e/MJ, but it has been adjusted by the energy economy ratio so can be considered a gCO2e/distance when comparing between fuel types.*

environmental impact than food that travels shorter distances (Rosenthal 2008). Shipping fresh apples and onions to Europe in the winter from New Zealand, for example, have lower emissions since locally produced options require months of refrigeration (Rosenthal 2008). In addition to seasonal efficiencies, production techniques have different impacts. Dairy products and lamb meat produced in New Zealand through well-managed grazing, then shipped to the United Kingdom, have been shown to have a significantly lower environmental impact than the same products grown in the United Kingdom (Saunders, Barber, & Taylor 2006).

Shifting consumption to foods that have a lower environmental burden is an additional consideration. More GHG's would be saved when less than one day per week's worth of calories moved from red meat and dairy products to chicken, fish, eggs, or a vegetable-based diet than buying all locally sourced food (Weber & Matthews 2008).

### 6.1.1 Refrigeration in transportation

Refrigeration is an important component of food transportation to reduce waste across the supply chain by maintaining the desired temperature of the product (not to cool it down; that is done before transportation). Refrigeration takes up about 40% of the total energy requirements during distribution (James & James 2014). The use of cooling equipment requires both energy and refrigerants. The energy needs are typically fueled by diesel or electricity and chemical refrigerants are the most common means of refrigeration (versus solid or liquid options such as carbon dioxide or nitrogen). Both of these are sources of GHG emissions. It has been estimated that 20% of the emissions come from direct refrigerant leaks and the rest from energy (Morawicki 2012). Hydrofluorocarbons (HFCs) are commonly used and have a global warming potential of 124-14800 carbon dioxide equivalent (Intergovernmental Panel on Climate Change [IPCC] 2007). Older equipment may still use chlorofluorocarbons (CFCs) and hydrochlorofluorocarbons (HCFCs), which are ozone-depleting (and thus have been banned from production) and have a global warming potential of 4750-14400 and 77-2310, respectively (IPCC 2007). Avoiding leaks is a critical means of preventing refrigerant loss and associated GHG emissions. Addressing refrigerant and energy impacts from refrigeration begins with preventative maintenance and regular monitoring. Ensuring the correct capacity of the chiller can help reduce energy demands from oversized equipment. There are also new technologies that have less GHG impact including hybrid and electric-only options for trucks. The Coca-Cola Company began including fully electric refrigerated trucks in its fleet in 2013 (O'Connor 2013). Electric chillers are used instead of the typical diesel-run units. Electric motors for refrigeration equipment are widely available for ships and rail. Solar-powered refrigeration and other lower GHG technology options are becoming more available.

### 6.1.2 EPA SmartWay

SmartWay Transport is a program of the U.S. Environmental Protection Agency (EPA) aimed at improving fuel efficiency and reducing GHG emissions and air pollution from the transportation supply chain industry. Since 2004 SmartWay Partners have saved $16.8 billion in fuel and eliminated 120.7 million barrels of oil with 51,800,000 million metric tons of carbon dioxide avoided (EPA 2014). The program has over 3,000 participating companies. These companies have committed to reporting performance on their freight operations and improving fuel efficiency. SmartWay also includes a recognition program for tractors and trailers that meet efficiency specifications. Each qualified tractor/trailer combination has the potential to save 2,000 to 4,000 gallons of diesel each year (EPA). Seeking out distribution partners involved in SmartWay or using SmartWay vehicles are potential ways to reduce environmental impacts from transportation.

## 6.2 Facility management

Distribution, retail, and food services all operate in facilities. These buildings use energy and water and generate waste. Operating facilities with an eye to conservation can save costs while reducing the environmental footprint. This is especially important since operation and maintenance costs throughout a building's life substantially exceed its initial design and construction costs (Public Technology Inc. 1996). The key elements for conservation-oriented, or green, building operation include the following (see Figure 6.3) (Baldwin 2012):

- Conserve energy and water
- Prevent, reduce, and avoid waste going to the landfill
- Clean green and optimize indoor environmental quality
- Purchase environmentally preferable products and services

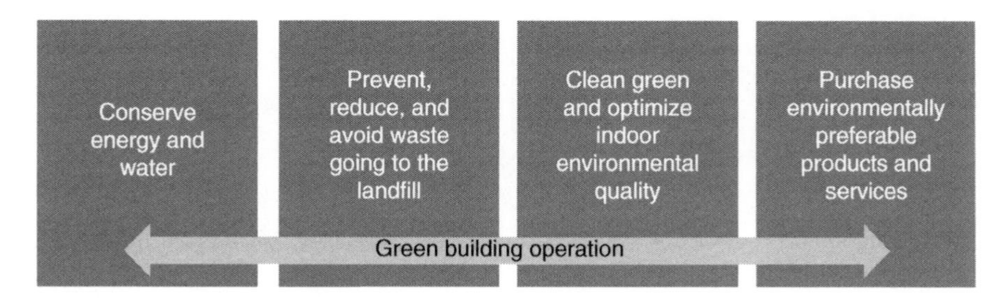


**Figure 6.3** Elements of green building operation (reproduced by permission of and adapted from Green Seal, Baldwin 2012).

Commonly implemented measures include:

- Preventative maintenance program
- Energy-efficient lighting
- Daylight and occupancy sensors for lighting
- Temperature control (programmed thermostat, turning down HVAC needs during low occupancy)
- On-off schedule for equipment (manual or automated)
- Energy-efficient equipment (Energy Star rated in the United States)
- Water-efficient fixtures and faucets
- Recycling program
- Green cleaning products
- Reusable cleaning tools
- Smoke-free facility

Energy use is universal in distribution, retail, and food service facilities. Electricity and heat are used for temperature control, lighting, storage, preparation, cleaning, and general maintenance and come with GHG emissions (as described in previous chapters). The amount of energy used varies depending on the functions of the facility. Greater energy is used per square foot in grocery stores and restaurants than warehouses or distribution centers (see Figure 6.4).

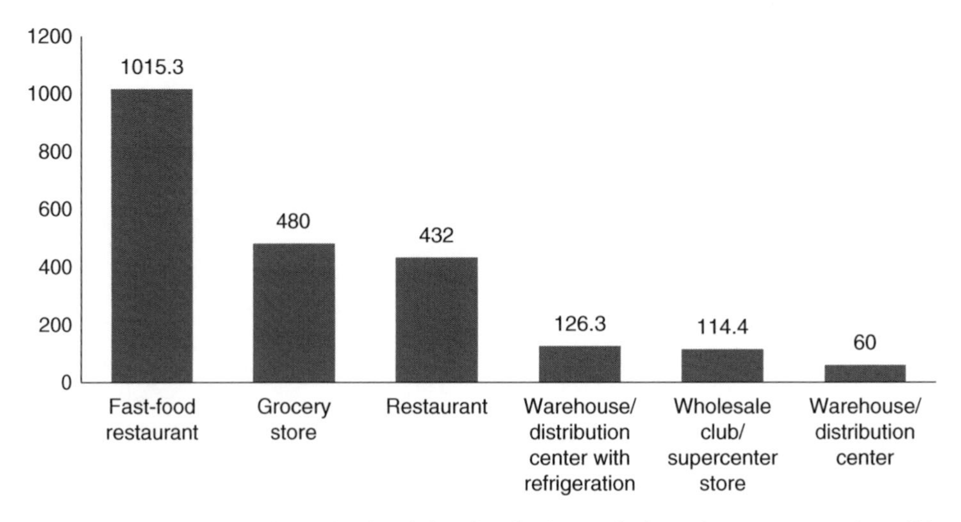


**Figure 6.4** Energy use intensity (EUI) for distribution and channels—source EUI kBtu/ft$^2$ (Energy Star 2013).

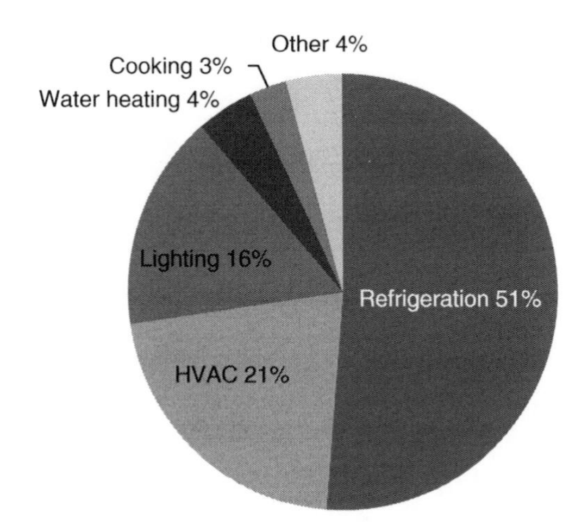


**Figure 6.5** Energy use in food retail (adapted from Davies & Konisky 2000).

Energy use is dominated by refrigeration which comprises over half the energy use at food retail outlets (see Figure 6.5) (Davies & Konisky 2000). The other main sources of energy use are lighting and heating/air-conditioning of the building. The result is that food retailers comprise 4% of all commercial building energy use in the United States (DOE g). Efficient use of refrigeration should be a priority of food retailers. Large savings can be gained by adding doors to freezer and chilled displays but ensuring use of the most efficient options is another critical strategy (James & James 2014). Using energy-efficient lighting in displays is also an opportunity (James & James 2014). Food services should focus more on kitchen efficiencies since food preparation requires the most energy in the facility (see Figure 6.6). Energy-efficient kitchen appliances and equipment are widely available on the market. The United States Energy Star program rates energy-efficient food service equipment to help companies find the best options. Addressing the other areas of the facility also offer opportunities to save costs and reduce environmental impact, remembering that careful tracking and monitoring alone can uncover notable savings. Warehouses and distribution centers may use less overall energy than other types of facilities but there is still room to optimize lighting, refrigeration, and battery charging efficiencies.

Water use depends on the facility type. Very little water is used in warehouse settings but more is used in food services. Food service operations use water for drinking, cooking, cleaning, sanitary uses, condenser cooling, and landscape irrigation. Water use is typically 3,000 to 7,000 gallons a day for a fast-food chain operation (Baldwin 2012). Irrigation can require significant amounts of water, around 10–50% of this total (Baldwin 2012). Taking away irrigation, water use is divided about equally between kitchen and sanitary uses with dishwashing and restroom uses predominating. Condenser cooling

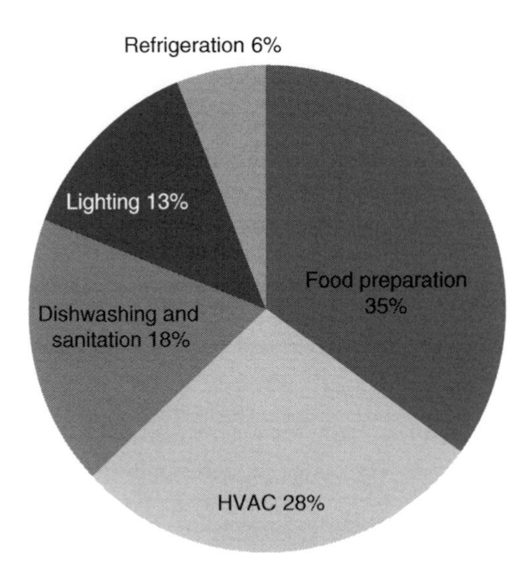


**Figure 6.6** Example of energy use in full-service restaurant (Energy Star 2009).

(when you combine the equipment that use condensers, such as ice makers, refrigerators, freezers, and HVAC) can also use as much water as dishwashing and restroom use if older technologies are used.

Waste comes from products, packaging, and supplies. Reducing, reusing, and recycling are key strategies to divert waste from the landfill to save on fees and make a better use of the material (packaging waste management strategies were covered in a previous chapter). Packaging comprises the majority of waste for most food retailers. Packaging is in the form of cardboard, paper, plastic, steel, aluminum, glass, and wood (Davies & Konisky 2000). Packaging contributes 34 to 50% of food service waste (California Integrated Waste Management Board [CIWMB] 2006). Nearly all the waste (both food and solid wastes) at warehouses, distribution centers, grocery stores, and food services can be reduced and recycled. Regrettably, typical landfill diversion rates are often low, only 30 to 35% for food services for example (CIWMB 2006). The primary fate of waste ends up being the landfill or incineration. When food waste breaks down in landfills it releases GHGs, such as methane, contributing to climate change. Landfills are one of the top methane emitters in the United States, responsible for 17% of methane emissions (EPA 2010a). Waste incineration also releases GHGs. As a result, waste reduction has significant potential for decreasing GHG emissions. Food waste management is discussed in another chapter that highlights the critical need to reduce food waste. When food waste does get generated it needs to be fully utilized by donating it to feed hungry people whenever possible. Additional strategies to avoid the landfill should then be pursued. The best approach is to avoid the production of waste to begin with. Conducting

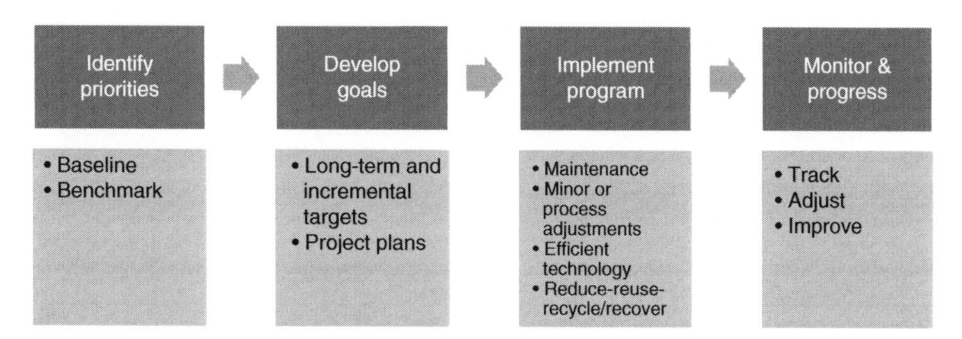


**Figure 6.7** Sustainability management process.

waste audits and walk-throughs help identify ways to reduce waste generation and better waste management approaches. Initial investment may be needed to address some issues but come with a payback. In warehouses and distribution centers, for example, investing in mechanical shrink-wrapping reduces waste and cost of the packaging material and provides work efficiencies and staff satisfaction.

Energy, GHG, water, and waste management approaches for warehouse, retail, and food service facilities are similar to those described in other chapters and generally follows the process of: identify priorities, develop goals, implement program, and monitor progress to continuously improve (see Figure 6.7). A thorough understanding of the amount and source of energy and water use and GHG emissions and waste is a critical starting point. The processing chapter described this effort, including how to select and develop a baseline. The baseline is used to compare progress against and determine appropriate goals and plans. Baseline development is best supplemented with site audits and competitor benchmarking to develop well-informed targets and plans. Three to ten year goals of 10 to 15% reductions are common. Considering needed reductions based on global targets of GHG emissions and water conservation is another way to set goals, called science-based goals. Stretch goals for renewable energy use and waste diversion should also be considered and are becoming more common. Walmart aspires to use 100% renewable energy and divert all of its waste, for example (Walmart b).

Implementation of the sustainability program at the facility level is driven by a foundation of maintenance and reduction efforts (see Tables 6.3 and 6.4 for energy and water). A sample of general building and facility approaches include (Baldwin 2012):

- Implement a facilitywide preventive maintenance program for equipment.
- Inspect and periodically replace or clean filters of air handlers and related equipment in accordance with manufacturers' recommendations.
- Evaluate heating, cooling, and ventilation capacity to see if such capacity matches up with occupancy and equipment loads.

**Table 6.3** The range of energy efficiency opportunities (adapted from Masanet, Therkelsen, & Worrell 2012)

| Minimal Investment ⟶ | | | Larger Investment |
|---|---|---|---|
| **Maintenance** | **Minor Adjustments** | **Operation Adjustments** | **Efficient Technology** |
| • Clean and tune equipment<br>• Calibrate temperatures, fill refrigerants<br>• Fix leaks<br>• Turn off idle equipment and lighting<br>• Keep cooler, refrigerator, and freezer doors closed | • Add timers and controls to lighting and equipment<br>• Update lighting<br>• Replace seals and curtains for coolers, refrigerators, and freezers<br>• Close monitoring and submetering<br>• Program thermostats and keep use of HVAC minimal | • Reduce downtime<br>• Adjust practices to reduce energy use (packing of refrigerators and freezers)<br>• For warehouses and distribution centers, conserve forklift battery use to minimize charging needs | • Upgrade older equipment<br>• Upgrade cooler and freezer walk-in doors to add high-speed closures or automatic openers/closers<br>• Generate heat and electricity on-site (e.g., solar panels, digesters) |

**Table 6.4** Water efficiency options (adapted from California Department of Water Resources)

| Minimal Investment ⟶ | | | Larger Investment |
|---|---|---|---|
| **Maintenance** | **Minor Adjustments** | **Operation Adjustments** | **Efficient Technology** |
| • Adjust equipment (water pressure)<br>• Fix leaks<br>• Turn off idle equipment<br>• Use dry cleaning before washing | • Low-flow fixtures<br>• Motion sensor, foot-pedal sinks<br>• Reuse water, rainwater collection<br>• Close monitoring and submetering | • Adjust practices to reduce water use<br>• Minimize irrigation for landscaping | • Upgrade older equipment |

- Establish temperature and humidity set-points in accordance with occupancy patterns, scheduling, and outside climate and seasonal variances.
- Periodically test, adjust, and balance the ventilation system to achieve proper airflow and air distribution to all zones and occupants within the building.
- Provide shade to your building during the warm months with trees and plantings and use reflective coatings on roofs and exterior walls.
- Install energy-efficient lighting and equipment.

- Reduce water flow rate and subsequent waste by installing low-flow, water-efficient faucets and other fixtures and appliances.
- Reduce outdoor water use on grounds with native plantings and reducing grass coverage.
- Minimize evaporation during irrigation by watering in the morning, using soaker hoses, and shortening watering times to avoid runoff.
- Capture and use rainwater and graywater as practical, when approved by local and state health and environmental authorities.
- Regularly track waste to quickly address sources.
- Ensure correct storage temperatures for food.
- Order, stock, and prepare the amount of food needed using detailed forecasting and planning.
- Reduce packaging coming in by working with suppliers.
- Recycle all packaging waste.
- Educate staff about energy and water conservation and waste reduction goals.

### 6.2.1 Refrigerants

Refrigeration is common at warehouse, retail, and food service facilities. Food service and food retail operations represent the largest commercial users of refrigeration in the United States, at 23.6% and 39%, respectively (DOE g). Most refrigeration equipment requires chemical refrigerants to help with cooling. Traditionally, chlorofluorocarbons (CFC) and other CFC-based refrigerants had been used. CFCs damage the protective ozone layer in the atmosphere so the sun's radiation penetrates to the earth's surface harming the environment and human health (e.g., skin cancer, cataracts, and immune system changes). The EPA regulations issued under Sections 601-607 of the Clean Air Act phased out the production and import of ozone-depleting substances, consistent with the schedule developed under the Montreal Protocol. However, facilities may have CFC refrigerants in older units. Another drawback of CFCs is that they are potent GHGs. As a result, CFC-based refrigerants should be phased out. The replacements most used are hydrofluorocarbons (HFCs) since they do not damage the ozone layer, such as CFCs, but they have the drawback of still being potent GHGs. About half of food retailer emissions come from HFCs (Environmental Investigation Agency [EIA]). As a result, it is best to move to refrigerants with lower GHG impacts such as hydrocarbons, carbon dioxide, ammonia, water, and air (sometimes referred to as natural refrigerants) which can all be used as cooling agents in refrigerators and freezers.

Retailers are beginning to take action to address the emissions from refrigerants. Carrefour prioritized moving away from HFC to reduce their GHG emissions (Fleury 2011). They have faced challenges in some climates, but have managed to reduce HFC use with hybrid approaches (HFC and carbon dioxide) while a 100% replacement solution is developed (Fleury 2011).

The EPA established the GreenChill Advanced Refrigeration Partnership in 2007 to encourage and recognize retailers that are adopting technologies, strategies, and practices that reduce emissions of ozone-depleting substances and GHGs and increase refrigeration system energy efficiency. The participants in the program have a refrigerant leak rate of less than 13%, compared to the 25% industry average, with additional efforts underway to further bring that down (EPA 2011). Refrigerant reductions from fewer leaks and other measures were responsible for the reduced emissions. Emissions are also being avoided by switching to more preferable refrigerants (EPA 2011).

### 6.2.2 Cleaning and indoor environmental quality management

The combination of effective cleaning, good housekeeping, and regular maintenance keeps the facility safe and running efficiently. This also enables food safety and good indoor air quality. Improving indoor air quality should not be overlooked since it supports the health and wellbeing of staff and customers. This may reduce illness and absenteeism and yield greater productivity, up to nearly 40 hours a year per staff member (Singh, Syal, Grady, & Korkmaz 2010).

A key starting point is continuous inspection of the building. The facility needs to be free of dust, debris, insects, standing water, and moisture damage or seepage. Be sure to not overlook areas around fresh air intakes, flues, vents, back-draft dampers, fans, and filters to minimize the spread of issues throughout the building. Fuel-burning appliances also need to be checked for leaks and proper venting. Table 6.5 outlines typical indoor air pollutants and means of addressing them.

Cleaning to address food safety concerns and regulations is critical. There are additional cleaning needs such as keeping the facility safe and healthy for its occupants (e.g., free of spills and dust/odors). However, the cleaning chemicals may contribute to the concerns and issues. For example, asthma is increasingly caused by use of cleaning products (Zoch Vizcaya, & Le Moual 2010). As a result, environmentally preferable cleaning products and procedures should be used. The following are key elements for an effective high performance cleaning approach (Baldwin 2012):

1. Standard procedures—when, how, and with what each surface gets cleaned.
2. Use of environmentally preferable products, tools, and equipment in good repair.

**Table 6.5** Methods to reduce common indoor air pollutants (EPA 2008)

| Pollutant | Reduction Methods |
|---|---|
| **Smoke** | Prohibit smoking indoors |
| **Combustion pollutants** | Vent appropriately and inspect fuel-burning appliances regularly for leaks and make repairs when necessary |
| **Radon** | Test for radon and install mitigation if necessary |
| **Organic compounds** | Reduce chemical use, use chemical products and furnishings with low VOCs, ventilate appropriately during product use, store chemicals appropriately |
| **Biological pollutants** | Clean regularly, ventilate appropriately, control humidity, remove excess moisture and standing water, change air filters appropriately |

3. Proper chemical use and handling—to minimize human exposure to cleaning chemicals.

4. Quality control to ensure that cleaning is meeting facility needs.

## 6.2.3 Environmentally preferable purchasing

A final component of running facilities is environmentally preferable purchasing (EPP). The primary focus should be on stocking the facility, store, or food service with more sustainable food (as described in previous chapters). EPP should also include supplies that are selected for a reduced effect on human health and the environment when compared with competing products or services that serve the same purpose (EPA 2010b). The benefits of purchasing these products include source reduction, fewer toxic chemicals, lower volatile organic chemicals, fewer GHG emissions, and reduced human health impacts, among others.

To help prioritize EPP efforts, it is sometimes useful to look at the amount of money spent on items and begin with the areas with the highest spend (Baldwin 2012). In addition, consider the overall environmental impact of the purchases. The bigger issues should be addressed. When making purchases look for as many of the applicable environmentally preferable attributes included in Table 6.6 as possible (Baldwin 2012). While some of these attributes are more preferable than others for particular products, the list can be scanned quickly when evaluating different purchases. For example, tissue paper products such as facial tissue and bathroom tissue should contain 100% recycled material content and as much post-consumer content available (e.g., at least 20% post-consumer material for bathroom tissue, 10% post-consumer material for facial tissue). Selecting products with these characteristics saves trees and related environmental burdens. Grocery bags should take a similar path with reduced weight and maximizing recycled and post-consumer content.

**Table 6.6** Common environmentally preferable attributes of products, beyond food purchases (reproduced by permission of and adapted from green seal, Baldwin 2012)

| Six Leading Environmentally Preferable Attributes |
|---|
| • Source-reduced, including lighter-weight, recycled, and post-consumer content, and salvaged and reclaimed |
| • Renewable materials that are organically or responsibly produced (e.g., FSC) |
| • Low VOCs and emissions |
| • Low toxins |
| • Energy and water efficient |
| • Reusable or recyclable |

EPP includes:

- Seeking out environmentally preferable attributes in products and services. Look for verification of these attributes from third-party certification programs.
- Experimenting with new environmentally responsible products and services.
- Incorporating environmental preferences into purchasing documents and discussions with suppliers and vendors.

### 6.2.4 Construction

Facility management at one time or another includes construction including repairs, renovations, or new construction. If the construction integrates more sustainable considerations at the onset greater energy efficiency and other benefits can be gained. Green buildings require one-third less energy, on average, compared to traditional buildings, delivering cost savings (Katz 2009). Green buildings also provide an environment where staff is healthier and more productive, delivering additional financial benefits. In the end, overall financial benefits of green buildings are estimated to be over ten-times the average initial investment (Katz 2003).

The larger the scale of the construction project, the more structured the effort needs to be. Table 6.7 outlines the flow of a construction project and how to integrate green goals, selection, planning, and implementation into the project. Initiating greening in the predesign phase helps assure the successful implementation of the goals for the project. Commissioning is included since it helps ensure that the green goals are met through a quality assurance process that verifies, on-site, that the building components are working correctly and that the plans are implemented with the greatest efficiency (Whole Building Design Guide [WBDG] 2010). Commissioning also provides training on how to maintain this level of performance.

**Table 6.7** Green construction process (reproduced by permission of Public Technology Inc. 1996)

| Construction Stage | Tasks |
|---|---|
| Predesign | Establish project goals and green design criteria<br>Set priorities<br>Establish budget<br>Assemble green team<br>Develop project schedule<br>Review laws and standards<br>Conduct research on green solutions<br>Select site |
| Design | Confirm green design criteria<br>Develop green solutions<br>Test green solutions<br>Select green solutions<br>Check cost |
| Bid | Clarify green solutions<br>Establish cost<br>Sign contract |
| Construction | Oversee green solution installation<br>Review substitutions and submittals for green products<br>Build project<br>Commission the systems<br>• Testing<br>• Operations and maintenance manuals<br>• Training |
| Occupancy | Recommission the systems<br>Perform green building operation<br>Conduct post-occupancy evaluation |

A green construction team helps run the project and typically includes the business owner/manager, architects, consultants, general contractor, and key subcontractors. A special green building consultant is especially helpful to include on the team when taking on a green construction project for the first time. The cost of adding this consultant for a restaurant project, for example, is between $15,000 to $40,000 and often represents the largest incremental cost for the green construction (Baldwin 2012). However, this cost is quickly offset with savings during the construction and especially during the operational phase of the project.

Green building criteria and frameworks are available to help guide the decision-making process for construction projects. These programs address considerations such as site selection, energy and water conservation, materials and resources, and indoor environmental quality. The most common frameworks in the United States are the United States Green Building Council's (USGBC) Leadership in Energy and Environmental Design (LEED) and the Green Building Initiative's (GBI) Green Globes rating systems. There is also an

International Green Construction Code (IGCC) from the International Code Council, Inc. These rating and code systems are particularly useful in helping provide guidance to the green construction team on priority areas. There are a number of resources that are available from the rating and code systems to help provide information on how to translate the criteria into actions. Many businesses find it useful to challenge themselves to meet as many of the provisions in the rating systems as possible, even if they are not pursuing a certification or recognition from the program. Common green building considerations to address include (Baldwin 2012):

- Efficient HVAC
- Energy-efficient appliances
- Daylighting and occupancy lighting controls
- Efficient lighting (usually LED)
- Water-efficient fixtures and faucets
- Recycled content furniture
- Recycling of construction materials

## 6.3 Gordon Food Service distribution and facilities improvements

Gordon Food Service is a distributor in North America. The company has a sustainability program with a focus on their operations including their fleet and facilities (Gordon Food Service [GFS]).

The company's fleet has a regular maintenance program to ensure that it meets all federal and local emissions standards. Gordon Food Service uses disciplined routing and load-building practices to reduce total miles driven and fuel consumption. Regular monitoring, enabled by technology, helps review and optimize performance and routing opportunities. The company also works with suppliers to find efficiencies with inbound freight. This is primarily achieved by focusing on sourcing products close to distribution centers and seeking out backhaul opportunities to limit overall supply chain miles and the associated carbon footprint. Gordon Food Service is also actively reviewing alternative fuel options for its fleet.

The company aims to run conservation-oriented facilities by integrating the USGBC LEED criteria into design and operations (GFS). Facility audits also identify key opportunities and plans for continuous improvement. After such an audit, a facility in Canada upgraded its lighting by moving from metal halide and sodium lighting fixtures to T5 fluorescent high bay lighting (BC Hydro

2011). Motion sensors were also added to turn off after 30 minutes. The energy savings from these changes cut electricity costs by 9%, with a nine-month payoff (BS Hydro 2011). The working environment was also improved with higher quality light (BC Hydro 2011). Old freezer doors were replaced with high-speed rolling doors with effective sealing to reduce energy use and had a payback of about 1.2 years (BC Hydro 2011).

Gordon also works to improve water use, waste handling, cleaning, and environmentally preferable purchasing. Several distribution centers utilize retention ponds to capture site runoff and use this water for irrigation purposes, when possible. The company has programs to increase plastic and paper recycling. Chemical-free treatment processes are used to treat water used in cooling applications. The company is also committed to increasing use of recycled-content material in packaging and other materials. As a result, Gordon Food Service is making important contributions to reduce the environmental burden from its distribution and facilities.

## 6.4 Food retailer J. Sainsbury addressing the environment

J. Sainsbury's sustainability program touches many considerations, including an extensive approach to improve transportation and store operations.

Beginning with driver engagement and better vehicle aerodynamics, Sainsbury's experienced an 8% reduction in km per liter and 3.8% reduction in distance since 2005/06 (J. Sainsbury). Route planning, scheduling, and real-time tracking also contributed to the improvements (Paragon b). This control led to a 15% reduction in turnaround time and a 12% improvement in empty running levels, helping cut GHG emissions (Paragon b). Alternative fuel use in the fleet is growing including 58 dual-fuel tractors that run on a biomethane and liquefied natural gas blend (J. Sainsbury). This fuel has 25% lower GHG emissions compared to diesel (J. Sainsbury). Sainsbury's is also testing a carbon dioxide refrigerant for trailers, for a further GHG reduction in the fleet.

Sainsbury's has already achieved significant energy and carbon emissions reductions in its facilities. However, the company has set an ambitious goal to reduce operational carbon emissions by 30% absolute and 65% relative by 2020, compared with 2005 (J. Sainsbury). The absolute goal is notable because many companies normalize their targets to adjust for business growth. Energy conservation is a critical program for the company. Lighting, which accounts for 20% of a Sainsbury's store's energy, is shifting to LEDs for existing and new stores. The company opened a store with 100% LED lighting and had a 59% energy savings over traditional lighting needs (J. Sainsbury). Customer feedback has been very positive about the lighting in the stores where LEDs are used.

Sainsbury's utilizes multiple approaches to reduce carbon emissions from energy use. Biomass heating with FSC-certified wood pellets provides 1.4% of demand for store heat and hot water. Individual sites have been able to generate up to 14 to 25% of the their energy demands from solar panels (J. Sainsbury). Additional options continue to be explored, such as ground source heat pumps that use refrigeration waste heat and deep underground cooling. However, the largest source of renewable energy comes from power purchase agreements where there are long-term contracts to use renewable energy generated by another party (typically on-site or nearby). This not only provides Sainsbury's low GHG and renewable energy but also stabilizes the price for energy. Already 13.5% of the company's electricity is provided through wind, biomass, and anaerobic digestion sources (J. Sainsbury).

The company has been advancing water efficiency at its facilities. By 2013, Sainsbury achieved a 50% relative reduction in water use compared to 2005/6 (J. Sainsbury). The company aggressively addresses leaks and seeks out opportunities to switch to efficient fixtures. Rainwater harvesting is included in all new stores and is being brought into existing stores. Sainsbury's has "water neutral" stores. To the company this means that 70% of the sites' needs are met with rainwater harvesting and other water efficiency efforts and the rest of the water demands are "offset" by sponsoring water savings initiatives in the community (J. Sainsbury).

Sainsbury's has a goal to put all waste to positive use by 2020. Addressing food waste is a critical share of the efforts toward achieving the goal. Sainsbury's works with the Waste and Resources Action Programme (WRAP) in the United Kingdom to reduce store and consumer food waste. The food waste that remains is handled such that the most value is derived from it, such as through donations, animal feed, and digestion. Sainsbury's already diverts all of its waste from the landfill; this includes food as well as solid waste such as packaging by recycling all cardboard and plastic.

All of Sainsbury's distribution sites have converted to non-ozone-depleting refrigerants by using ammonia and HFCs. Store refrigerants have been a target for GHG reduction, aiming to switch to natural refrigerants by 2030 (J. Sainsbury). The company already has 166 stores using carbon dioxide refrigerants (J. Sainsbury).

Sainsbury's environmentally preferable purchasing program effectively addresses food in the stores. The company also offers reusable shopping bags and has improved the standard free bags by reducing the weight by 9% since 2008 and moving to include 50% recycled material (J. Sainsbury).

Significant fuel, energy, water, and waste reductions demonstrate Sainsbury's position as a leader in sustainability. These achievements also help the company meet business objectives to cut costs, improve efficiencies, and satisfy customers.

## 6.5 Subway restaurants showing how to green operations

Subway restaurants provide a fresh quick service option to consumers across 100 countries. The company has committed to sustainability through enhancing sourcing and supply chain management and reducing energy, water, and waste.

With over 41,000 restaurants across the globe transportation is an important operation for the company. Subway looked strategically at its distribution and in addition to optimizing routes and loads, the company reviewed opportunities to move the location of its restaurants and distribution centers. The company reduced transportation, saving 17,000 shipments annually and reducing 21.8 million miles each year by moving facility locations (Subway).

Subway's conservation efforts extend to the restaurants with energy efficiency being a fundamental part of each site's operation. Lighting is a key source of energy use for the company. This is because the menu is less reliant on cooking and hot preparation, being more sandwich and salad-focused. All interior lighting has been switched to high efficiency lighting and the company continues to implement more improvements. The restaurants are using more LEDs in the store and for signage. The standard kitchen equipment in the store is also energy-efficient including Energy Star rated ice machines, refrigerated back counters, and reach in coolers and freezers (Subway).

Subway uses only low-flow faucets saving 182.1 million gallons of water annually (Subway). Motion sensors on hand sinks and dual flush toilets in many restaurants further reduce water demands. Some restaurants have taken an additional step by installing rain gardens, storm water pollutions prevention systems, native ground cover to reduce irrigation needs, and efficient landscape irrigation systems.

The company is aiming toward zero waste to landfill. This includes recycling and composting as much waste as possible, but it also includes reducing waste creation to begin with. The company works with its suppliers to minimize packaging coming into the restaurants and has reduced 6 million pounds of corrugate from sauce and tuna deliveries and 450,000 pounds of paper from cheese (Subway).

Subway restaurants are cleaned using environmentally preferable products. All of the cleaning chemicals, except for the sanitizer, are approved by a third-party as environmentally preferable (e.g., Green Seal). In addition to being nontoxic and biodegradable, the cleaning products are concentrated to remove extra water. The efficient packaging of the cleaning chemicals avoid 177,000 pounds of plastic and 154,150 pounds of corrugate each year, in addition to reducing extra transportation fuel (Subway).

Subway extensively pursues sourcing environmentally preferable supplies. Service items such as napkins, salad bowls, and utensils include recycled content and can be recycled or composted (Subway). The towel and tissue products are made with 100% recycled material and up to 75% post-consumer content. The napkins are made with 100% recycled content and up to 40% post-consumer content. They also have a reduced chemical burden being processed chlorine-free and printed with soy or water-based inks. Salad bowls and trays and lids have 25% recycled material and can be recycled. The company has taken this approach across all of its supplies by seeking out environmentally preferable options whenever they are available - even the mops have recycled content.

Subway has pushed its conservation efforts a step further with new site construction. The company developed its own green standard that is similar to the USGBC LEED standard. The company built 14 green buildings following their standard to conserve energy and water and reduce waste. Some of the features of these Subway ECO-Restaurants include (Subway):

- High efficiency HVAC (Heating, ventilation and cooling) equipment
- Energy Star-rated equipment
- High-efficiency lighting program
- Low-flow faucets and low-flow or dual-flush toilets
- Motion sensor lighting controls in restrooms
- Day light sensors in guest areas
- LED interior and exterior signage
- Non-smoking environment
- Outside air monitoring of $CO_2$
- Indoor air quality management during construction
- Reuse of at least 30% of store furniture
- Forest Steward Council (FSC)-certified wood moldings
- Low VOC (volatile organic compounds) materials, paints, and adhesives
- Electrical sub metering and thermal comfort monitoring
- Certified green cleaning program
- Recycling and construction waste management program
- LEED/Eco-restaurant educational material
- Furniture products not manufactured with or contain ozone depleting substances.

Subway's leadership in green operations demonstrates the range of efforts possible to address the environmental burden across transportation and facility management.

## 6.6 Summary

Food distribution, retailing, and restaurants contribute to the overall environmental burden of the food supply. Effective conservation efforts have helped keep this contribution low while also providing cost-savings benefits. Leading organizations have found ways to push past just reducing impacts by generating renewable energy and contributing to expanding freshwater availability. Donating food waste and recycling packaging also help make the most from the resources already in the supply chain. But there remains more to gain with the connection retailers and food services have with the consumer. There is an important opportunity to show by example and to educate consumers about the importance of sustainability in the food supply. Empowering consumers with this knowledge will help to expand the overall benefit across the supply chain (Table 6.8).

**Table 6.8** Summary of the principles–practices–potential of sustainability for distribution and channels

| | |
|---|---|
| **Principle** | *Food and ingredients are efficiently delivered across the supply chain and to the consumer* |
| **Practices** | • Distribute efficiently and conserve fuel use through:<br>  • Lower GHG-emitting modes<br>  • Fuel-efficiency measures<br>  • Route changes<br>  • Alternative fuels<br>• Conserve, reuse, and provide processing energy and water through:<br>  • Maintenance<br>  • Minor adjustments and process adjustments<br>  • Efficient technology, renewable energy, and water restoration<br>• Reduce, reuse, and recycle waste<br>• Clean green and optimize indoor environmental quality<br>• Purchase environmentally preferable products and services<br>• Green construction and renovation<br>• Educate and connect consumer to the value of a sustainable food supply |
| **Potential** | • Generate renewable energy<br>• Replenish water<br>• Feed hungry<br>• Reduce production needs<br>• Close the resource loop<br>• Empower consumer sustainability |

# Resources

US EPA Energy Star http://www.energystar.gov/
US EPA GreenChill http://www2.epa.gov/greenchill
US EPA SmartWay http://www.epa.gov/smartway/
US Green Building Council: www.usgbc.org

# References

Baldwin, C. (2012). *Greening food and beverage services: A green seal guide to transforming the industry*. Lansing, MI: American Hotel and Lodging Educational Institute.

Baldwin, C., Kapur, A., & Wilberforce, N. (2010). Restaurant and food service life cycle assessment and development of a sustainability standard. *International Journal of Life Cycle Assessment*. doi: 10.1007/s11367-010-0234-x.

BC Hydro. (2011). *Gordon Food Service: "Eye-opening" energy savings*. Retrieved from http://www.bchydro.com/news/conservation/2011/gordon_food_services_success2011.html

California Air Resources Board. (2010). *Subchapter 10. Climate Change Article 4. Regulations to Achieve Greenhouse Gas Emission Reductions Subarticle 7. Low Carbon Fuel Standard.* Retrieved from www.arb.ca.gov/regact/2009/lcfs09/lcfscombofinal.pdf

California Department of Water Resources. *Water use efficiency ideas*. Retrieved from http://www.water.ca.gov/wateruseefficiency/docs/Food.pdf

California Integrated Waste Management Board (CIWMB). (2006). *Cascadia Consulting Group report to board on the targeted statewide waste characterization study: Waste disposal and diversion findings for selected industry groups.* Retrieved from http://www.calrecycle.ca.gov/Publications/Disposal/34106006.pdf

Canning, P., Charles, A., Huang, S., Polenske, K., & Waters, A. (2010). *Energy use in the U.S. food system.* Retrieved from http://web.mit.edu/dusp/dusp_extension_unsec/reports/polenske_ag_energy.pdf

Coca-Cola Company. *Advancing energy efficiency and climate protection.* Retrieved from http://assets.coca-colacompany.com/48/4e/9f9f56e44bbb88a20877ff5872fa/SR09_Energy_ClimateProtct_20_23.pdf

Davies, T., & Konisky, D. (2000). *Environmental implications of the foodservice and food retail industries.* Retrieved from http://rff.org/Documents/RFF-DP-00-11.pdf

Dean Foods. (2011). *Dean Foods Alta Dena Dairy expands southern California fleet with five Ryder natural gas trucks*. Retrieved from http://www.deanfoods.com/our-company/news-room/press-release.aspx?StoryID=1579955

Del Monte Foods. *Delivering products responsibly*. Retrieved from http://www.delmontefoods.com/cr/default.aspx?page=cr_distribution

Energy Star. (2009). *Energy Star guide for restaurants: Putting energy into profit* Retrieved from http://www.energystar.gov/ia/business/small_business/restaurants_guide.pdf

Energy Star. (2013). U.S. energy use intensity by property type. Retrieved from https://portfoliomanager.energystar.gov/pdf/reference/US%20National%20Median%20Table.pdf

Environmental Investigation Agency (EIA). *Supermarket report*. Retrieved from http://eia-global.org/PDF/USSUPERMARKETREPORT.pdf

Fleury, J. M. (2011). *Roll out and experience of natural refrigerants based technology at Carrefour*. Retrieved from http://www.atmo.org/presentations/files/85_Carrefour-the-rollout-and-experience-of-natural-refrigerants.pdf

Fraser Basin Council. *Paradise Island Foods*. Retrieved from http://www.e3fleet.com/profile_paradise.html

Gordon Food Service (GFS). *Minimize environmental impact*, Retrieved from http://www.gfs.com/en/about-us/stewardship/au-stewardship-environmentalimpact.page

Gossling, S., Garrod, B., Aall, C., Hille, J., & Peeters, P. (2010). Food management in tourism: Reducing tourism's carbon 'foodprint.' *Tourism Management*. doi:10.1016/j.tourman.2010.04.006.

Greene, D. (2011). Reducing greenhouse gas emissions from U.S. transportation. Retrieved from http://cta.ornl.gov/cta/Publications/Reports/Reducing_GHG_from_transportation%5B1%5D.pdf

Heller, M. C., & Keoleian, G. A, (2000). *Life cycle-based sustainability indicators for assessment of the U.S. food system.* Center for Sustainable Systems, University of Michigan, Report CSS00-04. Retrieved from http://css.snre.umich.edu/css_doc/CSS00-04.pdf

Intergovernmental Panel on Climate Change (IPCC). (2007). *IPCC fourth assessment report: Climate change 2007. Section TS2.5.* Retrieved from http://www.ipcc.ch/publications_and_data/ar4/wg1/en/tssts-2-5.html

J. Sainsbury. *Sainsbury's: 20 x 20 factsheet*. Retrieved from http://www.j-sainsbury.co.uk/media/1790293/CSR%20Factsheet%20Environment.pdf

James, S., & James, C. (2014). Sustainable cold chain. In B. K. Tiwari, T. Norton, & N.M. Holden (Eds.), *Sustainable Food Processing* (pp. 463–496). Ames, IA: Wiley-Blackwell.

Katz, G. (2003). *The costs and financial benefits of green buildings: A report to California's sustainable building task force.* Retrieved from www.usgbc.org/Docs/News/News477.pdf

Katz, G. (2009). *Greening our built world: Costs, benefits, and strategies.* Washington, DC: Island Press.

Kruschwitz, N. (2014). *Ocean Spray, Tropicana team up on shipping to reduce emissions.* Retrieved from http://www.greenbiz.com/blog/2014/01/09/ocean-spray-tropicana-join-forces-reduce-emissions?mkt_tok=3RkMMJWWfF9wsRolvqjPZKXonjHpfsX56%2BwlW6C2lMI%2F0ER3fOvrPUfGjI4CTsVrI%2BSLDwEYGJlv6SgFSLHEMa5qw7gMXRQ%3D

Linkins, J. (2010). Final Deepwater Horizon flow rate estimate is likely too low, which benefits BP. Retrieved from http://www.huffingtonpost.com/2010/08/06/final-deepwater-horizon-f_n_673880.html

Masanet, E., Therkelsen, P., & Worrell, E. (2012). *Energy efficiency improvement and cost-saving opportunities for the baking industry*. Retrieved from http://www.energystar.gov/buildings/sites/default/uploads/tools/Baking_Guide.pdf

Morawicki, R. (2012). *Handbook of sustainability for the food sciences*. Ames, IA: Wiley-Blackwell.

O'Connor, M. C. (2013). *Coca-Cola launches first electric refrigerated truck fleet*. Retrieved from http://www.greenbiz.com/blog/2013/09/19/coca-cola-launch-first-electric-refrigerated-truck-fleet

Paragon a. *Glanbia saves 16% in six months.* Retrieved from http://www.paragonrouting.com/us/case-studies/glanbia-saves-16-in-six-months

Paragon b. *Paragon transportation optimization helps grocery retailer cut costs*. Retrieved fromhttp://www.paragonrouting.com/us/case-studies/paragon-transportation-optimization-helps-grocery-retailer-cut-costs

Pirog, R., & Benjamin, A. (2003). *Checking the food odometer: Comparing food miles for local versus conventional produce sales to Iowa institutions.* Retrieved from http://www.leopold.iastate.edu/pubs/staff/files/food_travel072103.pdf

Public Technology Inc. (1996). *Sustainable building technical manual*. Retrieved from http://smartcommunities.ncat.org/pdf/sbt.pdf

Rosenthal, E. (2008). *Environmental cost of shipping groceries around the world*. Retrieved from http://www.nytimes.com/2008/04/26/business/worldbusiness/26food.html?pagewanted=all&_r=0

Saunders, C., Barber, A., & Taylor, G. (2006). *Food miles: Comparative energy/emissions performance of New Zealand's agriculture industry.* Research Unit Research Report 285. Retrieved from http://www.lincoln.ac.nz/Documents/2328_RR285_s13389.pdf

Schneider, U. A., & Smith, P. (2009). Energy intensities and greenhouse gas emissions in global agriculture. *Energy Efficiency* 2, 195–206. doi: 10.1023/A:1009728622410

Singh, A., Syal, M., Grady, S. C., & Korkmaz, S. (2010). Effects of green buildings on employee health and productivity. *American Journal of Public Health* 100(9), 1665–1668. doi: 10.2105/AJPH.2009.180687

Subway *Environmental leadership*. Retrieved from http://www.subway.com/subwayroot/about_us/Social_Responsibility/EnvironmentalLeadership.aspx

U.S. Department of Energy (DOE a). *Efficient driving behaviors to conserve fuel.* Retrieved from http://www.afdc.energy.gov/conserve/driving_behavior.html

U.S. Department of Energy (DOE b). *Techniques for drivers to conserve fuel.* Retrieved from http://www.afdc.energy.gov/conserve/behavior_techniques.html

U.S. Department of Energy (DOE c). *Onboard idle reduction equipment for heavy-duty trucks.* Retrieved from http://www.afdc.energy.gov/conserve/idle_reduction_onboard.html

U.S. Department of Energy (DOE d). *Truck stop electrification for heavy-duty trucks.* Retrieved from http://www.afdc.energy.gov/conserve/idle_reduction_electrification.html

U.S. Department of Energy (DOE e). *Vehicle parts and equipment to conserve fuel.* Retrieved from http://www.afdc.energy.gov/conserve/equipment.html

U.S. Department of Energy (DOE f). *Alternative fuels data center.* Retrieved from http://www.afdc.energy.gov/

U.S. Department of Energy (DOE g). *In commercial buildings.* Retrieved from http://www.eia.doe.gov/kids/energyfacts/uses/commercial.html

U.S. Department of Energy (DOE). (2012). *Coca-Cola charges forward with hybrid delivery trucks.* Retrieved from http://www.afdc.energy.gov/case/1162

U.S. Environmental Protection Agency (EPA). *SmartWay transport overview.* Retrieved from http://www.epa.gov/smartway/about/documents/basics/420f12064.pdf

U.S. Environmental Protection Agency (EPA). (2007). *Mobile source emissions: past, present, and future: pollutants.* Retrieved from http://www.epa.gov/oms/invntory/overview/pollutants/index.htm

U.S. Environmental Protection Agency (EPA). (2008). *Care for your air: A guide to indoor air quality.* Retrieved from http://www.epa.gov/iaq/pubs/careforyourair.html

U.S. Environmental Protection Agency (EPA). (2010a). *Emissions inventory 2010: Inventory of U.S. greenhouse gas emissions and sinks: 1990–2008.* Retrieved from http://www.epa.gov/climatechange/emissions/usinventoryreport.html

U.S. Environmental Protection Agency (EPA). (2010b). *Basic information.* Retrieved from http://www.epa.gov/epp/pubs/about/about.htm

U.S. Environmental Protection Agency (EPA). (2011). *GreenChill progress report.* Retrieved from http://www2.epa.gov/sites/production/files/documents/GreenChill_ProgressReport2011_09062012.pdf

U.S. Environmental Protection Agency (EPA). (2014). *SmartWay program highlights.* Retrieved from http://www.epa.gov/smartway/about/documents/basics/420f14003.pdf

Walmart. *Index overview.* Retrieved from http://www.walmartsustainabilityhub.com/app/answers/detail/a_id/268

Walmart (Walmart b). Environmental sustainability. Retrieved from http://corporate.walmart.com/global-responsibility/environmental-sustainability

Weber, C. L., & Matthews, H. S. (2008). Food-miles and the relative climate impacts of food choices in the United States. *Environmental Science and Technology* 10.1021/es702969f. Retrieved from http://psufoodscience.typepad.com/psu_food_science/files/es702969f.pdf

Whole Building Design Guide. (2010). *Building commissioning.* Retrieved from http://www.wbdg.org/project/buildingcomm.php

Zoch, J. P, Vizcaya, D., & Le Moual, N. (2010). Update on asthma and cleaners. *Current Opinion in Allergy and Clinical Immunology* 10:114–120.

# 7 Food Waste

*Principle: Food and ingredient waste and loss are prevented across the supply chain and what cannot be avoided is put to a positive use.*

One-third of food is lost or wasted worldwide (Gustavsson, Cederberg, Sonesson, van Otterdijk, & Meybec 2011). The massive wastage comes with costs to the environment, society, and the economy. Not the least of which is the added pressure put on meeting the food needs of the growing population. The 60% increase in food production required by 2050 could be decreased substantially with cuts in food wastage and also save some of the $750 trillion worldwide of wasted or lost food (Food and Agriculture Organization of the United Nations [FAO] 2013a). Food wastage is happening across the entire supply chain. Agricultural production, postharvest storage and handling, and consumer use are the primary sources of food loss and waste. However, progress is needed across the supply chain to better feed our population and reduce the overall burden from food production.

## 7.1 The impacts from wastage

With global food wastage weighing in at 1.3 billion tons per year globally (see Figure 7.1), the environment burden of food wastage is remarkable (Gustavsson et al. 2011). Food wastage comes from two main sources, waste and loss. Food waste, a subset of wastage, includes the discards that are appropriate for human consumption (FAO 2013a). The rest of food wastage, loss, is deteriorated food that is no longer fit for human consumption because of decreases in mass or nutritional quality, not expiration (FAO 2013a).

The FAO conducted a study in 2013 to estimate the carbon and water footprints, land use, and biodiversity impact of this wastage. Notably, even without accounting for greenhouse gas (GHG) emissions from land use changes for food production such as deforestation, the carbon footprint of food produced and not eaten is estimated to be 3.3 Gtonnes of carbon dioxide equivalent (FAO 2013a). The FAO equates this global GHG contribution to be the third

*The 10 Principles of Food Industry Sustainability*, First Edition. Cheryl J. Baldwin.

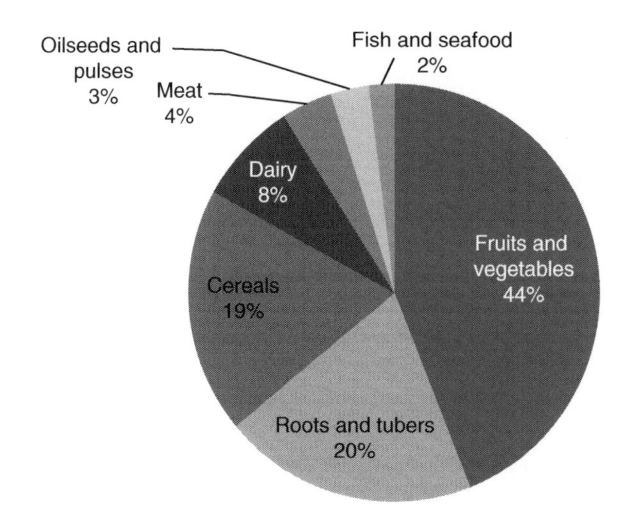


**Figure 7.1** Lost and wasted food by weight (reproduced by permission of World Resources Institute, Lipinski et al. 2013).

top emitter after the United States and China (FAO 2013a). The food wastage water footprint is estimated to be three times the volume of lake Geneva and to use close to 30% of the world's agricultural land area, with additional impacts such as biodiversity loss (FAO 2013a).

The top foods contributing to the carbon footprint of food wastage are cereals (34%), meat (21%), and vegetables (21%) (FAO 2013a). There are foods with a disproportionate impact to the carbon footprint, in relation to the amount of their waste, with cereals having substantial emissions from fertilizer and meat contributing significant emissions from animal production (FAO 2013a). The location of production plays a role with European-grown cereals not as carbon-intense as those grown in Asia (FAO 2013a).

Consumption is the largest contributor to the food wastage carbon footprint, at 37% (FAO 2013a). This is because so much carbon has been imbedded in the food through the life cycle by the time it reaches this point in the supply chain (FAO 2013a). For example, nearly half of the wastage carbon footprint at consumption is from the agricultural production stage (FAO 2013a).

Food wastage water and related impacts come mostly from cereals (52%) and fruits (18%), predominately from agricultural production (FAO 2013a). Meat and dairy products are largely (78%) responsible for land use from wastage (FAO 2013a). This is due to the land needed for grazing and feed production. However, grazing is mostly done on nonarable lands (FAO 2013a). Food wastage biodiversity loss is driven by agricultural production from land conversion and intensification (e.g., monoculture production) and as a result linked to crop production (70%).

Social effects from food wastage have been less studied. Indications suggest that they are as significant as the environmental impacts. A study on tomato wastage estimated to equate to 7 million working hours lost (Barilla Center for Food and Nutrition [BCFN] 2013).

These wide-ranging and dramatically destructive outcomes of food waste need to be resolutely addressed.

## 7.2 Reducing wastage

The European Commission has a 2020 target to reduce food loss and waste by 50%. The World Resources Institute (WRI) estimates that if this goal was met at a global scale that 22% of the food needs projected for 2050 could be filled (Lipinski, Hanson, Lomax, Kitinoja, Waite, & Searchinger 2013). The United Kingdom has already achieved a 21% reduction in household waste from 2007 to 2012 (Quested, Ingle, & Parry 2013) with additional progress well underway. The potential to achieve the needed cuts in food wastage appears to be within reach.

Agricultural production is the single largest source of wastage (see Figure 7.2). This is the situation across all parts of the world (FAO 2013a). Processing is consistently the smallest contributor to wastage. Wastage at the postharvest and consumption stages is more variable. Food losses in industrialized countries are as high as in developing countries but higher income regions had consumer wastage contributions of 31–39%, while lower income regions were at 4–16% (FAO 2013a, Gustavsson et al. 2011). Per capita food waste by consumers in Europe and North America is staggering, at 95 to 115 kg/year (see Figure 7.3 for what foods are typically wasted), while in

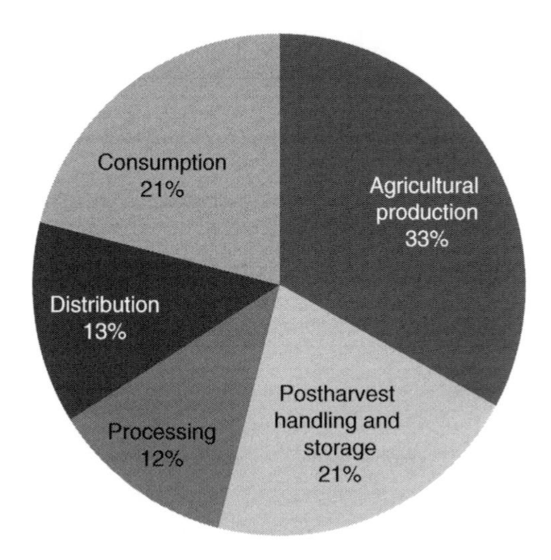


**Figure 7.2** Supply chain food wastage contributions (estimates based on FAO 2013a).

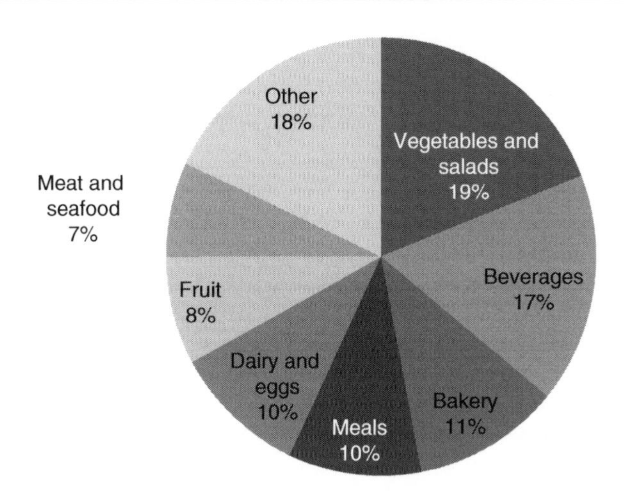


**Figure 7.3** Foods wasted by consumers (reproduced by permission of WRAP, Quested et al. 2013).

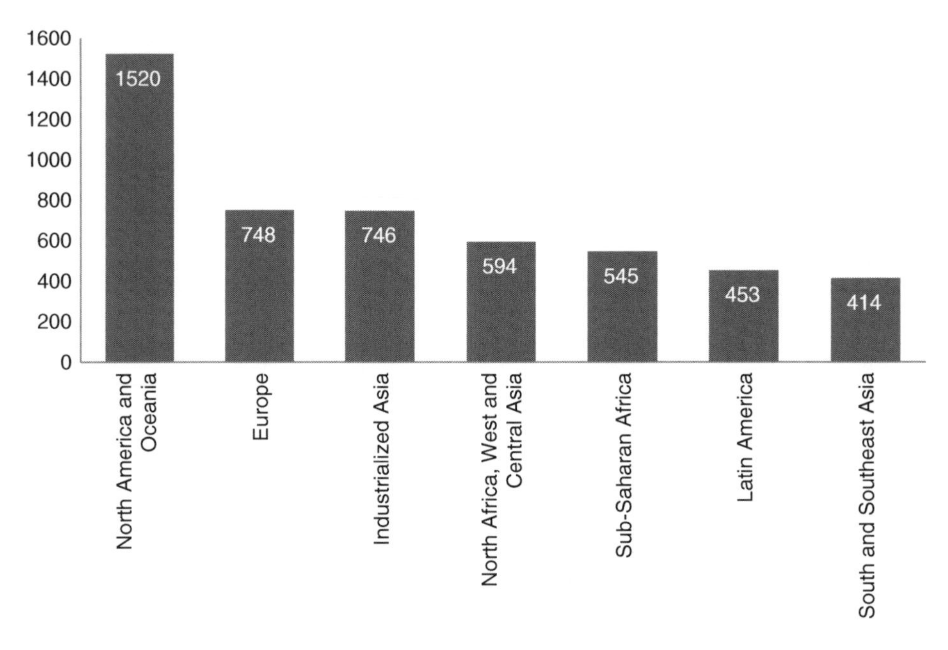


**Figure 7.4** Lost and wasted food by region, kcal/capita/day (reproduced by permission of World Resources Institute, Lipinski et al. 2013).

sub-Saharan Africa and South/Southeast Asia is only 6 to 11 kg/year (Gustavsson et al. 2011). This drives a significant impact from industrialized countries (see Figure 7.4).

Each stage of the supply chain has room for improvement (see Figure 7.5). The opportunities to address food wastage follow the hierarchy of reduce-reuse-recycle/recover-landfill (see Figure 7.6). Reducing food wastage is the most preferable approach since it avoids all impacts associated with the wastage. Source reduction

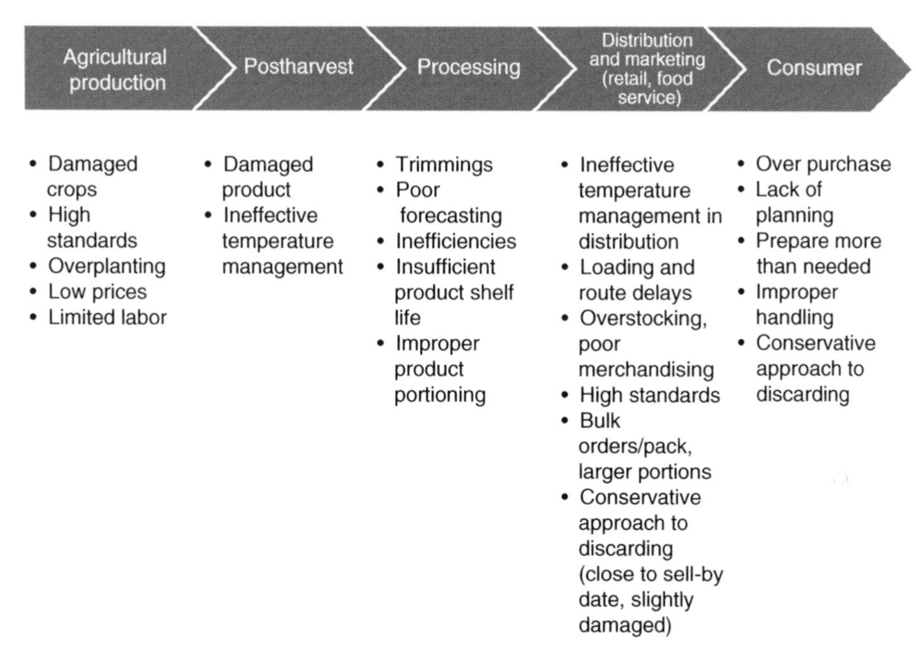


**Figure 7.5** Sources of food wastage food wastage across the supply chain.

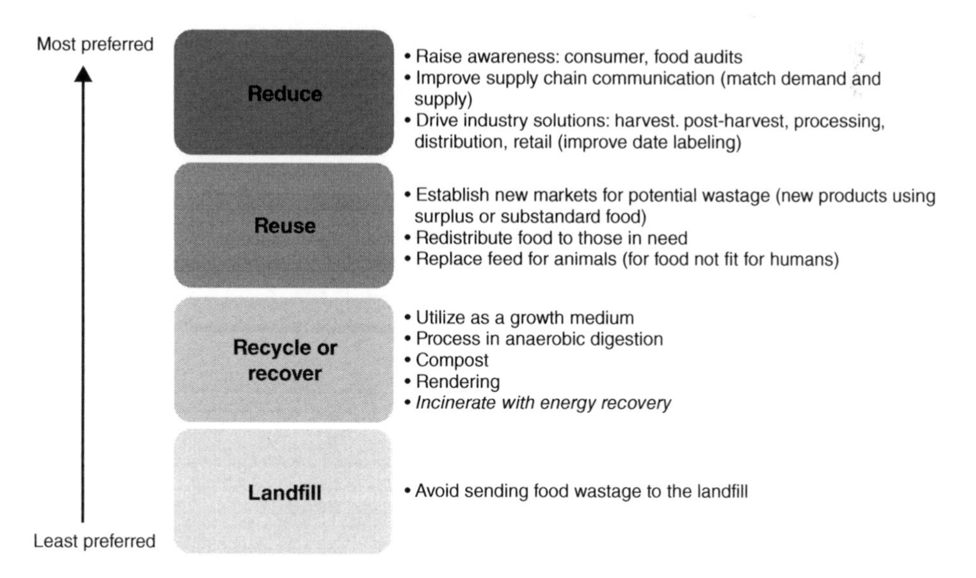


**Figure 7.6** Food wastage hierarchy.

is another term used for this aim. Simple measures can reduce waste including good management and planning. However, to address the root cause of some issues supply chain collaboration may be needed. As a result, there are collaborative-based organizations that are focused on addressing food wastage including the Waste and Resources Action Programme (WRAP) in the United Kingdom and the Food Waste Reduction Alliance (FWRA) in the United States. Partnering across the supply chain enables progress on issues that require industry alignment such as better matching demand and supply and standardizing date labeling on retail products.

The next most desirable strategy is to reuse the food for humans and secondarily for animals to replace other feed sources. Food wastage reused for humans can include repurposing the food to make a different product (e.g., sauce) or redistributing the food to those in need. Food donations can be made in nearly any region of the world and help hungry people. Logistics of the quantity accepted, timing of donation, and the ability of the receiving organization to pick up the food waste are sometimes hurdles for donations but can be overcome with discussing options with potential recipients and planning. Using food unfit for human use as animal feed is a viable option that is widely used. However, there can be some regulatory limitations on the type of food wastage used for animal feed (e.g., scraps from consumers may need heat treatment or may even be banned). The benefit with using food waste to feed animals is that it can replace diet options that have a greater environmental burden and at times can enhance the animal's performance. This has been observed when molasses, a by-product from sugar production, is fed to dairy cows. Increasing the proportion of molasses in the diet of milking dairy cows increases the milk yield, milk protein, and casein concentration and protein yield (Baldwin 2009). Another example is chicken eggshells providing a potentially useful livestock feed calcium enrichment ingredient as they are high in calcium with low levels of heavy metals (including lead and mercury), whereas the natural calcium carbonate source may be polluted with these elements (Faine 1995; Whithing 1994).

Recycling or recovering food wastage involves treating the discarded food for another purpose such as composting for a soil amendment, anaerobic digestion for energy, or rendering for tallow/meal. Precisely, recycling turns the food into another substance (e.g., compost) and recovering produces energy from the wastage (e.g., anaerobic digestion) (FAO 2013b). This stage in the hierarchy is much less desirable than the previous two since the inherent value is lost from the food, so most of the burdens of the wastage are not avoided. For example, one tonne of food wastage used for feeding pigs avoids 236 kg carbon dioxide equivalent (not including emissions avoided from land use change from feed production), whereas anaerobic digestion produces electricity (from methane and carbon dioxide) and heat to avoid 143 kg carbon dioxide equivalent (FAO 2013b).

Composting and digestion are becoming more common. They are both controlled decomposition processes to reuse the nutrients in the food wastage. Compost can be done using closed vessels or open environments; the best approach depends on the location, budget and resources, and amount of wastage. The final product of compost is used on fields to nourish the soil. Anaerobic digestion is a closed process that produces methane and carbon dioxide from fermenting the food. The methane is typically captured and used as an energy source. The remaining solids can be composted for use as a soil additive. Rendering of animal by-products to produce tallow or meals is commonly done but requires significant energy.

The final means of managing the food is to send it to the landfill. This is the least desired option since the food not only retains all of its burden but it has the additional impact of decomposition in the landfill emitting GHGs (methane). Food ending up in the landfill also comes with an added economic cost for disposal. Food products in landfills comprise about 12% of municipal solid waste in the United States and cost approximately $1.3 billion for their disposal (BCFN 2013).

The FAO details each of these approaches accordingly (FAO 2013b):

- Reduce. As the impact of food production on natural resources is enormous and increases while the food progresses on the food value chain, reducing food wastage is by far the best way of reducing the waste of natural resources. For example, if the supply-demand balance can be better adjusted on the front end, it means not using the natural resources to produce the food in the first place, thus avoiding pressure on natural resources, or using them for other purposes.

- Reuse. In the event a food surplus is produced, the best option is to keep it in the human food chain. This may call for finding secondary markets or donating it to feed vulnerable members of society, so that it conserves its original purpose and prevents the use of additional resources to grow more food. If the food is not fit for human consumption, the next best option is to divert it for livestock feed, thus conserving resources that would otherwise be used to produce commercial feedstuff.

- Recycle/Recover. The main recycling and recovering options are by-product recycling, anaerobic digestion, composting, incineration with energy recovery and rendering. All these options allow energy or nutrients to be recovered, thus representing a significant advantage over landfill.

- Landfill. Landfilling organic waste causes emission of gases such as methane (a very potent greenhouse gas) and potentially pollutes soil and water, let alone odor and other societal nuisance. Landfills should be the last resort option for food waste management, especially in a context of increased land scarcity for earth citizens.

This is slightly different from the U.S. Environmental Protection Agency's (EPA) food recovery hierarchy (see Figure 7.7) (EPA). The point of difference

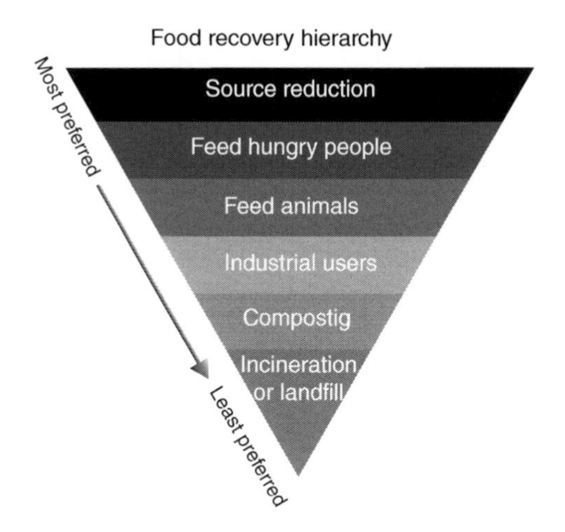


**Figure 7.7** Food waste hierarchy (EPA).

is that the EPA includes incineration with landfilling in the hierarchy because there is a very low energy value in burning food wastes and thus is considered a last resort option.

## 7.2.1 Waste management

The approach to effective waste management has a common process across the supply chain (see Figure 7.8). It begins with an internal focus that involves measuring the waste, setting goals and a plan, and taking action. It is important to track progress and adjust as needed to keep the program on target. The overall effort should also engage upstream and downstream partners to maximize its overall effectiveness. However, all of this should be done with the right focus, best determined by understanding the business case and the situation.

The first step is to understand how much waste there is, what the waste is, the source of the waste, and where the waste ends up (landfill, compost, digester, etc.). This is best done with an audit that establishes a baseline. The baseline, similar to others (e.g., GHG emissions), is used to develop goals and plan and track progress. The goals and plan should follow the waste hierarchy, reduce-reuse-recycle/recover-landfill (see Figure 7.6). The highest priority goals should be aimed to reduce or avoid waste then followed by diverting waste from the landfill and incinerator. An emerging goal for food processors is "zero waste." This typically means that at least 90% of waste is not sent to the landfill or incinerator. However, what is missing is a reduction target. Both of these have clear a business case since waste equates to lost profit. However, there are a number of approaches to achieve these goals. This is where it is important to assess the options and determine the best strategies

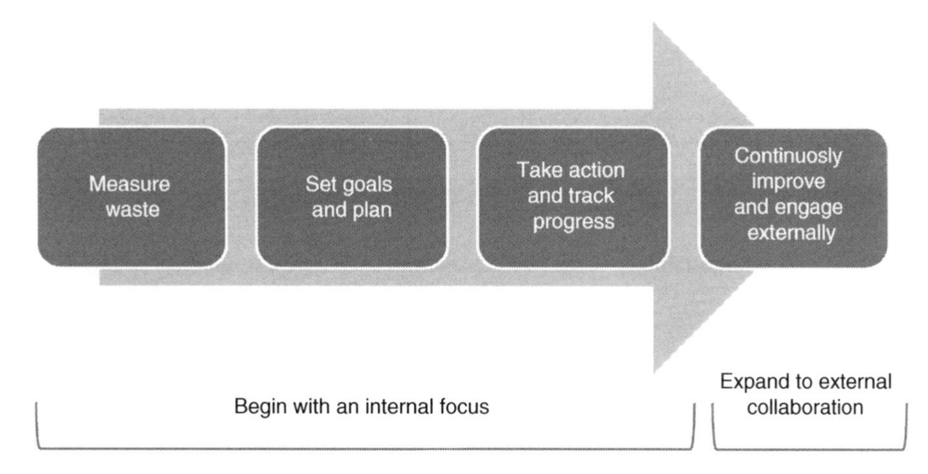


**Figure 7.8** General process to advance waste management efforts.

for your business (e.g., resources). Once the plan is established it is time to take action and track progress.

As organizations advance on these efforts it becomes clear that more progress can be made with external collaboration with upstream and downstream partners. There are collaborative organizations to enable this (e.g., FWRA, WRAP) and many companies work directly with their suppliers (Robertson 2013). Another opportunity lies with assisting the areas in the supply chain that have the greatest need. This may be further away from the business and harder to establish a business case. However, investing in postharvest improvements in developing countries, for example, provide value across the supply chain.

#### *7.2.1.1 Agricultural production and postharvest handling and storage*

The majority of food wastage occurs during agricultural production and postharvest handling and storage. At the farm this is due to factors such as marketing prices, labor availability, and the inherent risk of needing enough quality product at harvest time. Postharvest waste and loss typically arise from spoilage due to insufficient infrastructure for transportation, storage, cooling, and markets (FAO 2013b). This can be addressed with global access to proper handling and storage technologies, better ripening and sanitation management, widespread use of low GHG-emitting refrigerants, and improved infrastructure (e.g., roads) (Lipinski et al. 2013).

The inherent risks from agricultural production are a cause of the wastage at this stage. Risks including weather, pests, and price volatility are a regular part of farming. Producers often manage this by overplanting to ensure they have enough product that meets market standards at harvest time. This leads

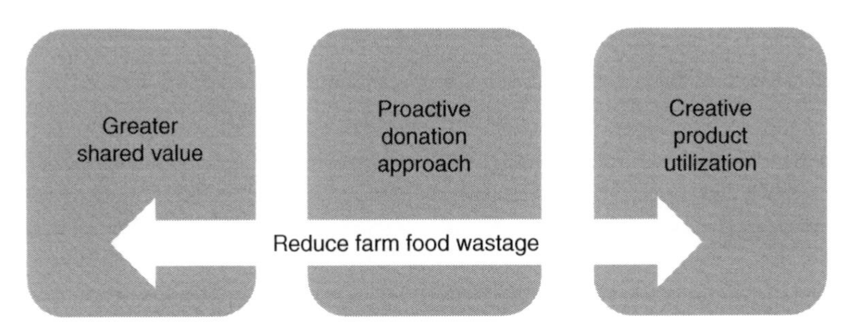


**Figure 7.9** Approach to reducing food wastage at the farm.

to wastage since more than is needed is grown (NRDC). This is worsened when a grower is selling their product for market prices rather than a preestablished price/contract since they experience greater price volatility. This volatility can lead to harvest costs being substantially more than the market price would provide for the product. The product is then left in the field. Another source of wastage on the farm is the lack of sufficient skilled labor for harvest—more and more there isn't enough labor to affordably harvest the product (NRDC).

Reducing food wastage at the farm can be achieved with greater shared value across the supply chain, proactive donation approaches, and creative product utilization (see Figure 7.9) (NRDC). Shared value may be most effectively achieved with longer-term relationships with customers (processors or retailers). The success of this approach lies in the assurance for continued business in the face of uncontrollable issues such as storms or pests, thus overcompensating for risks are less likely (NRDC). Greater farmer collaboration through cooperatives may also buffer some of the risk and enable more long-term relationships with customers (since some customers prefer a larger producer/supply pool) (NRDC). Use of the food in the field for donations could be increased with efficient harvesting and distribution techniques but requires a proactive approach. During the main harvest, for example, the off-standard product could be collected at the same time (concurrent picking) for donations (NRDC). A donation program has benefits beyond reducing food wastage; employees have greater job satisfaction and enjoy participation, there are usually tax credits, and the community gets the needed support and returns it with good public relations (NRDC). Additional market value may be possible by putting the product to new uses. The development of baby carrots produced from undersized or misshapen carrots were a notable market win, for example (NRDC). The options with new products may extend beyond the current customer; such opportunities may be with schools, universities, prisons or government institutions (NRDC).

Postharvest loss and waste is a substantial contributor to overall food wastage (21%) (Gustavsson et al. 2011). Developing countries have greater wastage at this stage than industrialized regions. The main factors leading to postharvest wastage include (Postharvest Education Foundation [PEF]):

- Knowledge and resource gaps (practices and technologies).
- Poor temperature control for stored food.
- Improper or overly strict sorting of product.
- Insufficient packaging to protect the food.

Despite significant wastage at postharvest there are opportunities to effectively address it. The challenge with reducing postharvest losses is that it requires some investment in training and technologies globally. Quick wins can be achieved with training on proper packing, handling, sorting, and pest control (BSR 2013). Greater investment can improve packaging and storage and transportation infrastructure (BSR 2013). The success of an FAO program demonstrates the potential at this stage where the organization built 45,000 metal silos that each support a single farmer and helped cut food loss to almost zero in the areas impacted (Lipinski 2013).

#### *7.2.1.2 Processing, distribution, and retail*

The stages between the farm and the consumer generate less food wastage than up or downstream. Despite the smaller contribution to wastage, many food processors are aiming for zero waste at their processing facilities (i.e., 90% diversion from landfill and incinerator including other sources of waste such as packaging). The benefits for these organizations and others that move in this direction are not just reduced waste and related impacts, but also cost savings.

Processing milk, chocolate/sugar confection, brewing/distillation, and meat products produce more waste than other food processes (Niranjan & Shilton 1994). This is due to a high level of trimmings (both edible and inedible) and processing inefficiencies (Gunders 2012). Companies may be losing 4% of product with inefficient packaging lines alone and improving overall process control on the production line typically reduce costs by 5% (Envirowise). Reengineering production processes and product designs and finding secondary uses for the food waste (e.g., within production, for anaerobic digestion) are additional opportunities to reduce waste and avoid the landfill. A combination of equipment maintenance and design changes help minimize waste (Envirowise). Industry respondents in the United States have implemented many of these approaches and divert 94.6% from the landfill (BSR 2013b). This includes sending any remaining food waste offsite for other valuable uses, primarily for animal feed (see Figure 7.10) (BSR 2013b). There is potential for moving up the food waste hierarchy by donating more for human use. Addressing food bank storage and transportation

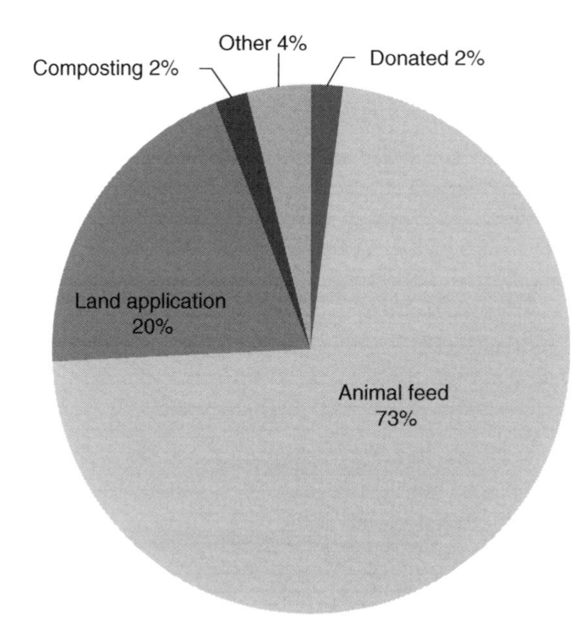


**Figure 7.10** Uses of diverted food waste from processors (reproduced by permission of Grocery Manufacturers Association and Food Marketing Institute, BSR 2013b).

limitations or costs for company storage and transportation may help improve this (BSR 2013b).

Food processors are able to reduce food waste downstream as well. Products can be designed to maximize shelf life (with limited reliance on synthetic additives) and for appropriate portioning for the consumer to reduce waste. This may involve product or packaging changes, or a combination of the two. The product label should aim to include clear storage instructions to help the consumer manage and extend the life of the product appropriately (e.g., freezing and defrosting instructions).

Distribution waste and loss occurs when temperature control is not properly managed or if a shipment cannot be sold or diverted for donation (Gunders 2012). Ensuring the refrigeration system is in good working order and set to the correct temperatures is important, as is planning loading and routes to allow for proper temperature (e.g., avoiding idling in hot days) (Gunders 2012).

The retail stage can be a source of wastage not just at the store, but also upstream and downstream. Extremely stringent product standards can drive waste upstream and selling products in large pack sizes can cause the consumer to put more food in the trash. At the store the following are all sources of food waste (Gunders 2012):

- Overstocking product displays to result in greater damage of product.
- Stocking only the best quality product and discarding the rest.

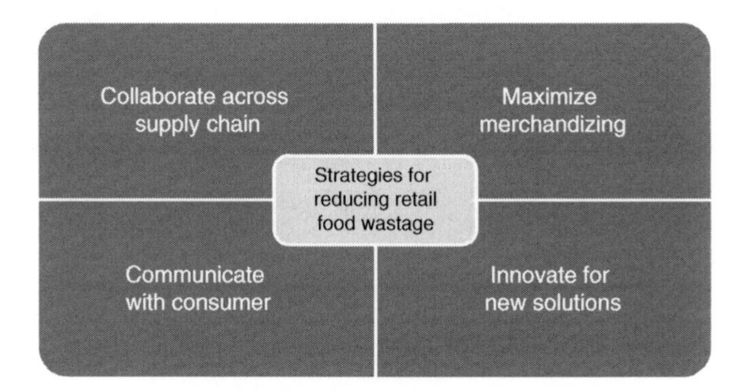


**Figure 7.11** Approaches to reduce retail food waste.

- Ordering quantities that are greater than needed.
- Keeping a wide selection of fresh, prepared foods on display at all times.
- Discarding product close to the "sell by" date, even if that date is not an accurate indication of quality.
- Removing damaged, outdated promotional products, and unpopular products from the shelf and not finding another use for them.

Retailers can address these wastage drivers to get more value from the product in the store (see Figure 7.11). Effective ordering is key to this by ensuring that the correct amount and quality is ordered. Another opportunity is to collaborate with upstream partners to design specifications that minimize wastage across the supply chain. There are merchandizing solutions to reduce food waste that are straightforward including selling product at a discount when it ends up close to its shelf life or is slightly damaged or outdated (Gunders 2012). There are new technologies being tested to extend product shelf life or reduce waste, such as a packaging strip that absorbs the ripening gas, ethylene so produce can last longer (Gunders 2012). At the retailer only 55.6% of the food waste being diverted, so there is significant room for improvement (see Figure 7.12 for uses of diverted food waste) (BSR 2013b).

Food services have a similar role as retailers, but have a unique interface with consumers since their food waste is often generated on-site (see Table 7.1 reasons for food waste). The food service can, in effect, reduce its own and the consumer's waste. Knowing the sources of the waste (with audits) and engaging staff to address the issues is the best starting point. Solutions then include training staff and having them assist with adjusting the menu (even limiting some choices or having flexible portions), merchandising (fewer items on display), and ingredient control (staging

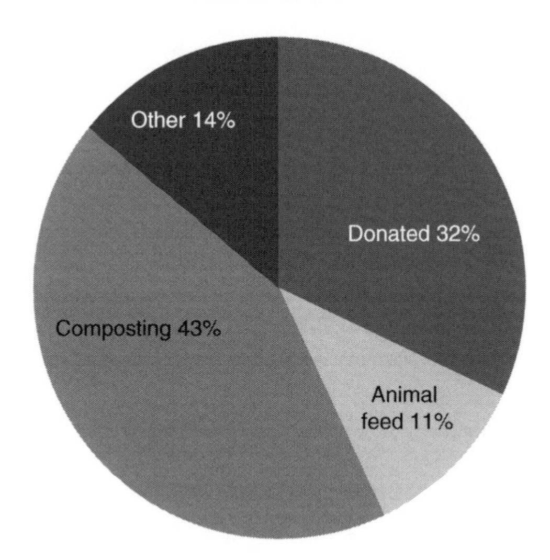


**Figure 7.12** Uses of diverted food waste from retailers and wholesalers (reproduced by permission of Grocery Manufacturers Association and Food Marketing Institute, BSR 2013b).

**Table 7.1** Sources of food service food waste (reproduced by permission of Green Seal, Baldwin 2012)

| Pre-Consumer Food Waste | Post-Consumer Food Waste |
|---|---|
| • Unknown demand and forecasting challenges (e.g., seasonality, weather patterns, and local competition influence the ability to forecast demand correctly)<br>• Desire not to run out<br>• Inefficient or inaccurate production procedures (e.g., large batch cooking regardless of demand)<br>• Poor staff communication to determine demand (e.g., front of house and back of house)<br>• Staff behavior (e.g., ignore forecast, use ingredients as packed and not what is needed)<br>• Unskilled trimming and preparation<br>• Prioritizing merchandising (e.g., display too much)<br>• Safety (e.g., address reoccurring discards due to safety) | • Large portion sizes<br>• Lax service models (e.g., self-serve environments such as cafeterias, grill buffets, and noncommercial food service operations such as college dining halls encourage patrons to take more food than they will consume)<br>• Menu acceptance |

ingredients, using the correct amount). Forecasting and purchasing the right amount of food is also important. Novel approaches have emerged in food services to address food waste including advanced tracking of waste, trayless university dining cafeterias, and restaurants with smaller portion sizes with notable success—achieving a 43% reduction in food wastage with computer tracking and a 30% reduction with trayless cafeterias (Gunders 2012).

#### 7.2.1.2.1 Kraft foods advancing zero waste processing

Global food manufacturer, Kraft Foods, has a focus on reducing its waste. The company has already achieved a notable 50% reduction of waste (per ton of production) since 2005 (Kraft 2012). Kraft continues to advance its efforts and has 36 manufacturing plants in 13 countries sending no waste to the landfill (Kraft 2012).

A critical element of the company's approach is including measurement and tracking of waste as a key "environmental performance indicator." Measurement and tracking began with establishing a waste baseline in 2008 along with a reduction goal (at the time it was a 15% reduction goal). To progress toward the goal, continuous improvement programs across the plants seek out engineering and operation's solutions—getting more efficiency. A plant in the Philippines reduced waste by one-third since 2008 by targeting critical points where waste was being created (Kraft 2012).

Kraft diverts food waste from the landfill using a variety of approaches. Two plants in the state of California sent more than 100 tons of food waste to farms for animal feed (Kraft 2012). A Philadelphia Cream Cheese plant in Beaver Dam, Wisconsin, worked with the local community to build an anaerobic digester that uses the dairy by-product whey to generate electricity for the local grid (Kraft 2012). A plant in Vienna, Austria, sent 250 tons of used coffee bean husks to a biomass power plant to create electricity and heat for the community (Kraft 2012).

With significant internal progress, Kraft has an opportunity to expand waste reduction beyond its own operations. The company has joined the collaborative supply chain effort lead by WRAP in the United Kingdom to address waste. This kind of external collaboration can help Kraft contribute additional and substantial wins across the supply chain.

### *7.2.1.3 Consumption*

Food wasted by the consumer is a leading source of total food wastage. The waste at this stage is occurring primarily in industrialized countries, 80% of this waste (FAO 2013a; Lipinski et al. 2013). Consumer food waste comes with a greater environmental burden since it is further down the supply chain, accumulating the impacts from each stage along the way. The reasons for this waste are consumer behavior and ineffective supply chain communication (FAO 2013a). Consumers often purchase more than is needed (impulse or bulk purchased), don't plan effectively to lead to purchased items not being used or preparing more than is needed, improperly store or handle the food to speed up spoilage, and are conservative in determining when to discard food (Gustavsson et al. 2011; Quested et al. 2013). Consumers may also not value the impact or cost of wasted food. The supply chain is another source of consumer waste such as when there are only bulk sizes available for purchase or when quality standards are too restrictive, based on size or aesthetics (FAO 2013a).

The foods wasted by consumers in the United Kingdom come from all categories of foods (see Figure 7.3) (Quested et al. 2013). The top ten food types thrown away were, by weight, were as follows (Quested et al. 2013):

1. Standard bread
2. Fresh potatoes
3. Milk
4. Meals (homemade and preprepared)
5. Carbonated soft drinks
6. Fruit juice and smoothies
7. Poultry meat
8. Pork meat
9. Cakes
10. Processed potatoes (e.g., chips)

Efforts to improve consumer education can address many of the root causes for food waste (see Figure 7.13). The product code dates are a source of confusion and can cause food wastage. In-depth research from the United Kingdom determined that "sell by" dates should be avoided and that consumers could use more information about how to extend the product's usefulness, e.g., "freeze by" dates (WRAP 2013a). These approaches combined with extensive consumer education and supply chain collaboration has proven to reduce waste in the United Kingdom—food and beverages thrown away that could have been consumed fell by 21% between 2007 and 2012 (Quested et al. 2013).

#### *7.2.1.4 WRAP leading food wastage reduction across the supply chain*

The United Kingdom government program Waste and Resources Action Programme (WRAP) is an innovator in advancing solutions that address food wastage. WRAP partners with the supply chain and stakeholders in

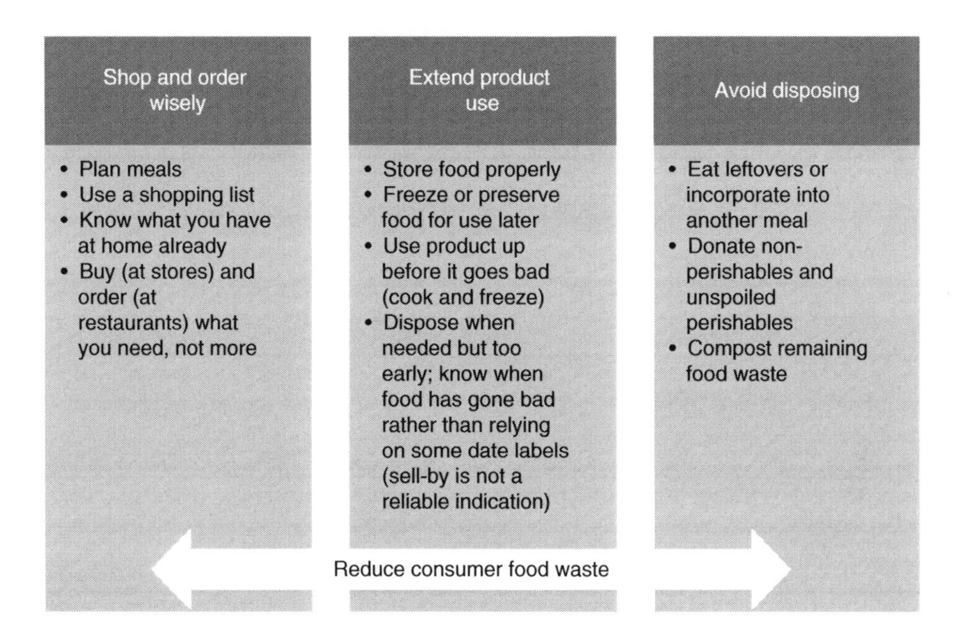


**Figure 7.13** Approaches to reduce consumer food waste.

collaborative efforts that have been extremely successful. This includes programs that have led to the consumer food waste drop of 21% noted earlier as well as a 7.4% reduction in food and packaging waste in retail between 2009 and 2012 (WRAP 2013b).

WRAP coordinates programs that address supply chain and consumer waste. Both of these programs are informed by WRAP's detailed study that uncovers the amount, source, and reasons for the waste. This evidence-based research has paved the path forward globally on food waste issues. WRAP's work to identify the reasons for consumer waste is particularly noteworthy. This insight led to further investigation on packaging and product date labeling. As a result of this work there is a global discussion about removing "display until" dates and adding labeling about proper storage (e.g., freeze) to provide more clear communication to consumers.

WRAP's research, guidance, tools, case studies, and expertise supports the supply chain in waste reduction efforts. A cornerstone program is the "Courtauld Commitment," a voluntary agreement aimed at reducing waste across the supply chain. Signatories of the commitment have contributed to a reduction in food waste by improving products, addressing processing inefficiencies, and helping raise consumer awareness about food waste reduction practices. Food retailer in the United Kingdom, Asda (owned by Walmart), increased the shelf life of over 1,500 chilled products by an average of one day by cutting down delivery road miles and using more simplified store systems to get product to the shelf faster (WRAP 2013c). Carlsberg UK addressed process inefficiencies such as adding a ramp to avoid 3 tonnes of hop extract loss and modifying the malt process to reduce and capture waste for animal feed, diverting 260 tonnes of waste each year (WRAP 2013c). Retailer Warburtons removed the confusing display date label from their products so consumers only saw the "best before" date (WRAP 2013c).

WRAP spearheads extensive and effective consumer awareness campaigns on food waste reduction. Understanding that consumers are driven by the costs associated with wasted food led to ads drawing this link, "food lovers save money," for example (BSR 2013c). A blog titled, "The easy way to save £50 a month," explained steps for consumers to reduce food waste by freezing food and effective as much as possible to portion control (WRAP 2013c). This theme carries through the consumer website, www.lovefoodhatewaste.com.

WRAP's leading-edge programs continue to cut food waste across the supply chain and doing this with a collaborative approach for a broader impact globally.

#### *7.2.1.5 Tesco tackles food waste*

Tesco, a leading food retailer has taken a strong position on food waste. The company estimates that uneaten food costs families about £700 a year (Johnston 2013). So the company deployed innovative approaches to reduce waste at the store, in consumer's homes, and through supply chain partners.

Tesco bases its efforts on an understanding of food waste—where the waste occurs, the reasons for the waste, and what is needed to address it. The company measured waste across their supply chain for 25 of their top-selling foods. This helped define the need to work not only at its stores, but also upstream with agricultural production and the supply chain as well as downstream with consumers. Additional measurement techniques are in development to be able to consistently measure waste at stores across all markets and track progress over time (Tesco a).

At the store level, Tesco is focusing on addressing hotspots of bakery and produce. Most parts of the store have less than 1% of the total food waste in the supply chain coming from the retail sector. Bakery and produce have notably higher waste at retail. These two departments represent over 60% of the store's food waste (Tesco b). As a result, bakery displays were changed to have less bread out which also reduced the amount needing to be prepared unnecessarily (Tesco b). The company coupled the bakery improvements with updated IT systems to optimize stock ordering and daily production planning (Tesco b). Produce displays were also improved including adding banana hammocks that reduce product damage (Tesco b). To expand the implementation of effective approaches, Tesco developed a "blueprint of best practices" for all of their markets (Tesco b). For a team in Thailand, store-level changes included training staff to minimize product damage and improving forecasting and store ordering (Tesco b).

Upstream efforts reach directly to farms. This connection is helping identify where product specifications may be resulting in edible products going to waste and possible ways to address this to reduce waste (Tesco a). A strong connection in the supply chain is also enabling better forecasting for suppliers and producers to reduce surplus (Tesco a). Production, storage, and transportation opportunities are also being assessed. Two techniques for grapes include trialing new varieties with a longer life and using new techniques to protect grapes in rainy geographies (Tesco b).

Tesco is working to provide informative and consistent messaging about storage on the package, online, and in customer communications for produce to help reduce consumer waste (Tesco b). Promotions are also being optimized to address hotspots. Tesco will not offer multi-buys on larger packs and is exploring "intelligent" promotions that allow customers to mix and match products (Tesco b). The dates on the product are being clarified to reduce confusion by removing "display until" dates.

Tesco is paving the way for less waste in the food supply to become the norm. Along the way they are establishing new ways to communicate across the supply chain, up and downstream, to address this important issue.

## 7.3 Summary

The FAO estimates that cutting food wastage by just 25% can feed the current hungry population. This also helps address the increasing food demands of the growing population. These significant societal benefits should be enough to drive action, but the additional benefits of reduced risks, environmental burden, and economic costs should help tip the balance. A number of efforts underway have demonstrated each of these benefits, pushing closer to the aim of avoiding waste and loss. With the leaders showing the way, there is sure to be more to follow and food wastage will see dramatic reductions (Table 7.2).

**Table 7.2** Summary of the principles-practices-potential of sustainability for food wastage

| | |
|---|---|
| **Principle** | *Food and ingredient waste and loss are prevented across the supply chain and what cannot be avoided is put to a positive use.* |
| **Practices** | • Prioritize food wastage management to:<br>  • Reduce<br>  • Reuse<br>  • Recycle/recover |
| **Potential** | • Feed hungry<br>• Reduce production needs<br>• Conserve resources |

## Resources

United Nations Food and Agriculture Organization (FAO) http://www.fao.org/save-food/en/
Waste and Resources Action Programme (WRAP) http://www.wrap.org.uk/
U.S. Environmental Protection Agency http://www.epa.gov/smm/food recovery/

## References

Baldwin, C. (Ed.). (2009). *Sustainability in the food industry*. Ames, IA: Wiley-Blackwell.

Baldwin, C. (2012). *Greening food and beverages services: A green seal guide to transforming the industry*. Lansing, MI: American Hotel and Lodging Educational Institute.

Barilla Center for Food and Nutrition (BCFN). (2013). *Combating waste: defeating the paradox of food waste*. Retrieved from http://www.barillacfn.com/wp-content/uploads/2013/06/BCFN_Magazine_CombatingWaste.pdf

BSR. (2013). *Losses in the field: An opportunity ripe for harvesting*. Retrieved from http://www.bsr.org/reports/BSR_Upstream_Food_Loss.pdf

BSR. (2013b). *Analysis of U.S. food waste among food manufacturers, retailers, and wholesales*. Retrieved from http://www.foodwastealliance.org/wp-content/uploads/2013/06/FWRA_BSR_Tier2_FINAL.pdf

BSR. (2013c). *Household food waste: Opportunities for companies to provide solutions*. Retrieved from http://www.bsr.org/reports/BSR_Reducing_Household_Food_Waste.pdf

Envirowise. Food and drink essentials: Environmental information for the industry.

Faine, M. P. (1995). Dietary factors related to preservation of oral and skeletal bone mass in women. *J. Prosthet. Dent*, 73, 65–72.

Food and Agriculture Organization of the United Nations (FAO) (2013a) *Food wastage footprint: impacts on natural resources summary report*. Retrieved from http://www.fao.org/docrep/018/i3347e/i3347e.pdf

Food and Agriculture Organization of the United Nations (FAO) (2013b) *Toolkit: reducing the food wastage footprint*. Retrieved from http://www.fao.org/docrep/018/i3342e/i3342e.pdf

Gunders, D. (2012) *Wasted: How American is losing up to 40% of its food from farm to fork to landfill*. Retrieved from http://www.thinkeatsave.org/docs/NRDC_Wasted_FINAL_20120827.pdf

Gustavsson, J., Cederberg, C., Sonesson, U., van Otterdijk, R., & Meybeck, A. (2011). *Global food losses and food waste: Extent, causes and prevention* (from the Food and Agriculture Organization of the United Nations). Retrieved from http://www.fao.org/docrep/014/mb060e/mb060e00.pdf

Johnston, I. (2013). *Tesco vows to act after study confirms huge food waste*. Retrieved from http://www.independent.co.uk/news/uk/home-news/tesco-vows-to-act-after-study-confirms-huge-food-waste-8893015.html

Kraft Foods. (2012). Kraft Foods waste-reduction sustainability success stories. Retrieved from http://www.kraftfoodscompany.com/SiteCollectionDocuments/corp/waste_reductions_factsheet.pdf

Lipinski, B. (2013). *10 ways to cut global food loss and waste*. Retrieved from http://www.wri.org/blog/10-ways-cut-global-food-loss-and-waste

Lipinski, B., Hanson, C., Lomax, J., Kitinoja, L., Waite, R., & Searchinger, T. (2013). *Reducing food loss and waste*. Retrieved from http://www.unep.org/pdf/WRI-reducing_food_loss_and_waste.pdf

Natural Resources Defense Council (NRDC). *Left-out: An investigation of the causes and quantities of crop shrink*. Retrieved from http://docs.nrdc.org/health/files/hea_12121201a.pdf

Niranjan, K., & Shilton, N. C. (1994). Food processing wastes—their characteristics and an assessment of processing options. In E. L. Gaden (Ed.), *Environmentally responsible food processing.* New York: American Institute of Chemical Engineers.

Postharvest Education Foundation. *Food waste or postharvest food losses?* Retrieved from http://postharvest.org/is_it_food_waste_or_postharvest_loss0.aspx

Quested, T., Ingle, R., & Parry, A. (2013). *Household food and drink waste in the United Kingdom 2012.* Retrieved from http://www.wrap.org.uk/sites/files/wrap/hhfdw-2012-main.pdf

Robertson, K. (2013). *A 4-step approach to reducing corporate food waste.* Retrieved from http://www.greenbiz.com/blog/2013/08/13/4-step-approach-reducing-food-waste

Tesco a. *Reducing food waste: Our approach.* Retrieved from http://www.tescoplc.com/index.asp?pageid=594#tabcontent

Tesco b. *Reducing food waste.* Retrieved from http://www.tescoplc.com/assets/files/cms/Food_Waste.pdf

U.S. Environmental Protection Agency (EPA). *Food recovery challenge.* Retrieved from http://www.epa.gov/smm/foodrecovery/

Waste and Resources Action Programme (WRAP). (2013a). *Consumer insight: Date labels and storage guidance.* Retrieved from http://www.wrap.org.uk/sites/files/wrap/Technical_report_dates.pdf

Waste and Resources Action Programme (WRAP). (2013b). *Courtauld Commitment phase 2 helps delivers £3.1bn in savings*, Retrieved from http://www.wrap.org.uk/content/courtauld-commitment-phase-2-helps-delivers-%C2%A331bn-savings

Waste and Resources Action Programme (WRAP). (2013c). *Courtauld Commitment 2: signature case studies.* Retrieved from http://www.wrap.org.uk/sites/files/wrap/CC2%20case%20studies%20-%20Dec%202013.pdf

Whithing, S. J. (1994). Safety of some calcium supplements questioned. *Nutr. Rev.*, 52, 95–105.

# 8 Nutrition, Security, and Equity

*Principle: Safe and highly nutritious food is accessible and affordable to promote and support a healthy population.*

*Principle: Producer equity and rural economy and development are strengthened with fair and responsible production and sourcing.*

*Principle: Safe and suitable working conditions are provided to support employees across the supply chain.*

The food system is intended to provide safe and nutritious food to support a healthy population. Even though enough is being produced to feed the population, it is not yet reaching everyone equally (Food and Agriculture Organization of the United Nations [FAO] 2012a). On one end we are faced with overnutrition and diseases related to overconsumption, with over one billion adults overweight and another half a billion obese globally (FAO 2012a). Yet, on the other end, one billion are hungry and not getting enough food to eat (see Figure 8.1) (FAO 2012a). Shockingly, the people we rely on to produce our food are among the most susceptible to food insecurity. Producer access to resources, fair prices, and markets limit their ability to move out of poverty and meet their food and other needs. This threatens not only their livelihood and well-being but also the global food supply. All of the Principles covered in this book interconnect, but there are three important concepts that will be discussed together in this chapter: nutrition, food security, and producer equity and safety. Supporting the producers of our food will help build a more resilient supply that can feed the population. This includes small- and medium-sized farms and women in particular. These issues are often overlooked in the supply chain but are gaining the attention deserved and needed to increase food production and global development.

*The 10 Principles of Food Industry Sustainability*, First Edition. Cheryl J. Baldwin.

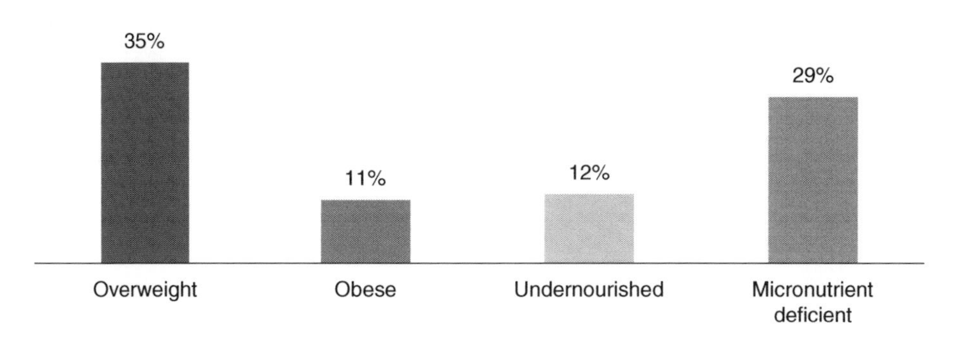


**Figure 8.1** Rates of over nutrition and under nutrition (reproduced by permission of CGIAR 2014).

# 8.1 Nutrition

Food safety and nutritional quality are essential elements of our food supply. Good nutrition is the foundation for human health and well-being and economic productivity (FAO 2013a). Nutrient-dense and diverse diets are key determinants of nutritional quality and outcomes (FAO 2013a). Good nutrition, however, begins with food that is safe for consumption. The safety concerns of the food supply include not only microbial contamination and related illnesses but also chemical and allergen considerations.

## 8.1.1 Food safety

Foodborne illnesses affect 48 million annually in the United States including 3,000 deaths each year (United States Centers for Disease Control and Prevention [CDC] 2012). The most prevalent foodborne pathogens causing these illnesses are overwhelmingly associated with animal products (CDC 2012). The following pathogens represent over 90% of foodborne illnesses (CDC 2012):

- Norovirus
- *Salmonella*, nontyphoidal
- *Clostridium perfringens*
- *Campylobacter* spp.
- *Staphylococcus aureus*

The CDC recommends, "People who want to reduce their risk of foodborne illness should assume raw chicken and other meat carry bacteria that can cause illness, and should not allow these foods to cross-contaminate surfaces and other foods. People should also cook chicken and other meat well, [and] avoid consuming unpasteurized milk and unpasteurized soft cheeses. It is always best to cook seafood thoroughly" (CDC 2013a). However, prevention upstream of consumers can address most issues. There are several good agricultural and manufacturing practices that can reduce the spread of microbes in livestock animals and prevent the contamination of foods (CDC 2012).

Addressing foodborne illness has been complicated by the increase of antibiotic-resistant bacteria (Mathew, Cissell, & Liamthong 2007). Such antibiotic resistance is growing and threatens human health. Each year two million Americans are infected with bacteria that are resistant to antibiotic treatment and at least 23,000 die from these infections (CDC 2013b). Resistance occurs when bacteria are exposed to antibiotics. Prudent use of antibiotics is critical to avoid overexposure and accelerating resistance development.

Agriculture accounts for 70% of antibiotics sold in the United States—with the pork industry emerging as one of the heaviest users, especially of antibiotics used to treat humans (Forini, Denison, Stiffler, Fitzgerald, & Goldburg 2005). Use of antibiotics for meat and poultry production has reached record levels in the United States and continues to increase (U.S. Food and Drug Administration [FDA] 2013a). Two classes of antibiotics that are commonly used in livestock and poultry are of medical importance to humans, penicillin and tetracyclines. As a result, the World Economic Forum and the World Health Organization have called for reduced use of such antibiotics in agriculture to help minimize the potential that the medically important medicines used to treat humans may become infective (World Economic Forum [WEF] and World Health Organization [WHO] 2012). Good management with proper sanitation and husbandry practices that keep the animals well-nourished and healthy (e.g., environment, weaning practices) are a first line of defense to help address this issue and are practical solutions worth optimizing.

Foodborne illness and health issues can also arise from chemicals in foods. These may stem from pesticide residues, packaging materials, processing residues, veterinary residues (hormones, antibiotics), and even intentionally added ingredients. Some artificial food additives and colors have a significant effect on the behavior of children with attention-deficit hyperactivity disorder (ADHD) (Schab & Trinh 2004). The antioxidant preservative butylated hydroxyanisole (BHA) and the bread making process aid potassium bromate are both classified as possible human carcinogens by the leading authority, the International Agency for Research on Cancer of the WHO (IARC) (IARC 2014). Partially hydrogenated oils (e.g., soy) contain trans-fatty acids that are linked to an increased incidence of coronary heart disease. The CDC in the United States stated that reduction of trans fat can prevent 7,000 deaths from heart disease and up to 20,000 heart attacks each year (FDA 2013b). Trans fat content of food must be labeled in the United States and the Food and Drug Administration is moving to ban their use because of the health issues with these ingredients (FDA 2013b).

Pesticide residues on produce can include highly hazardous chemicals, illegal residues, and high levels of chemicals. Leaf vegetables, apples, berries, grapes, celery, peaches, and potatoes often carry the highest residue levels, including illegal chemicals. Some pesticides used are highly hazardous and can lead to acute and severe illness. The two most commonly applied herbicides in the United States are suspected endocrine disruptors that can change reproductive function (Horrigan, Lawrence, & Walker 2002). Both the WHO and FAO recommend action to reduce and eventually eliminate the most highly hazardous pesticides (e.g., permethrin use/residues on lettuce or carbendazim and oxamyl use/residues on cucumbers, among others) (FAO; WHO 2009). Integrated Pest Management (IPM) is a leading alternative pest management strategy that is effective at reducing pesticide needs and residues. IPM saves the producer money, improves the environment, and protects health.

The Global Food Safety Initiative (GFSI) provides a benchmarking of food safety audit programs to enable harmonization across companies. Compliance to a GFSI audit is becoming a standard requirement for many customers (e.g., retailers). Some of the programs commonly used are Global GAP and Primus Labs. Each of these programs address fundamental control points to prevent pathogenic and chemical contamination. These programs also touch on pesticide safety by generally aligning with IPM.

Food allergies are a growing concern. In the United States, 15 million people have food allergies (Food Allergy Research and Education [FARE]). This includes a 50% jump, between 1997 and 2007, of children with food allergies (FARE). The main cause linked to the increase in allergies is excessive cleanliness, especially in developed countries, which interferes with the development of an effective immune system (University of California Los Angeles Food and Drug Allergy Care Center). Unlike microbial and chemical contamination, addressing food allergies is more about consumers avoiding certain foods than addressing supply chain issues. Exposure to trigger foods can result in typical allergic reactions and can become life-threatening such as, itching and swelling of the mouth, vomiting, diarrhea, cramps, hives, tightening of the throat and trouble breathing, and blood pressure drop. As a result, food companies in the United States are required to label the major food allergens on products through the Food Allergen Labeling and Consumer Protection Act and other countries have similar rules. The following major food allergens account for 90% of reactions (FDA 2009):

- Peanuts
- Tree nuts
- Fish
- Crustacean shellfish

- Milk
- Eggs
- Soybeans
- Wheat

#### 8.1.1.1 *Whole foods market moving away from concerning pesticides*

U.S. natural foods retailer Whole Foods Market (WFM) has an extensive sustainability program integrated into the core of their business. A new program that rates produce on production practices takes the company a step further. In 2014, the company began rating all produce to indicate the level of responsible practices used to grow the food (WFM 2013). Among the environmental categories rated are water, soil, biodiversity, waste, energy, and climate. Social considerations such as farm worker welfare are also included. Of particular interest is how pesticide use is rated. WFM is taking an industry-leading approach with an extensive prohibited list of pesticides and restricting additional pesticides. WFM is doing this with the aim to eliminate the most toxic pesticides from the food supply. The program also encourages growers to measure and reduce other pesticide use. This will help protect consumers, farm workers, and the environment.

### 8.1.2 Overnutrition and obesity

Overconsumption of food contributes to excess body weight and a host of related health issues. Overweight and obesity are associated with an increase in cardiovascular disease, hypertension, type-2 diabetes, stroke, certain cancers, musculoskeletal disorders, and some mental health conditions. The medical costs from obesity in the United States were $147 billion in 2008 with 35.7% of adults obese (CDC 2013c). What may have been considered an issue limited to affluent countries is now emerging to become the leading health concern in developing countries (United Nations Environment Programme [UNEP] 2012). Approximately 1.4 billion adults are overweight and nearly 500 million are obese across the globe (WHO 2012). These numbers are on a dramatic upward trend.

Overnutrition issues impact populations differently. In developing countries, as income increases there is a move to a more Western-style diet that is higher in fat, sweeteners, and animal products. Diet-related diabetes incidence is growing fastest among the more affluent that can afford the higher fat foods in the Middle East, India, and China (UNEP 2012). When China nearly doubled its meat consumption during the 1990s chronic diseases rapidly became a more common cause of death (Shelke, Van Wart, & Francis 2009). On the other hand, obesity is the highest among the lower-income population in the

United States (UNEP 2012). In developed countries, the poor are more impacted owing to limited access to cost-effective healthy options (CDC 2013d). In addition to consuming higher-fat foods, people in developed countries are consuming at least 20 to 30% too many calories (CBC 2013). Eating a healthier diet without overconsuming would help tackle obesity and its health issues and at the same time reduce the demands on the food production system (CBC 2013).

An added challenge in developed countries is that children are targeted with advertising for unhealthy foods. As a result, the ads viewed by children are almost exclusively for products that are high in fat, sugar, or sodium (Yale Rudd Center for Food Policy and Obesity [Rudd] 2012). The advertising techniques are persuasive including premium offers, promotional characters, the theme of taste, and the emotional appeal of fun (Jenkin, Madhvani, Signal, & Bowers 2014). This advertising is considered a key contributor to childhood obesity and unhealthy eating (Public Health Law Center [PHLC]). The Children's Food and Beverage Advertising Initiative was established in the United States by the Council of Better Business Bureau to address this concern. Although most major companies have committed to following it, ads to children remain focused on unhealthy food options (Rudd 2012). Companies need to look beyond this guidance and take a more responsible approach in how they communicate to children about food choices.

The World Health Organization (WHO) has a global strategy on diet, physical activity, and health that outlines the important role for the food industry in promoting healthy diets (WHO). The key concepts include the following (WHO 2013):

- Reducing the fat, sugar, and salt content of processed foods.
- Ensuring that healthy and nutritious choices are available and affordable to all consumers.
- Providing understandable and meaningful nutritional information to consumers.
- Practicing responsible marketing especially those aimed at children and teenagers.
- Ensuring the availability of healthy food choices and supporting regular physical activity in the workplace.

#### *8.1.2.1 General mills improving product nutrition*

Providing convenient and nutritious food is embedded in the mission of General Mills: "Nourishing Lives." To deliver on this mission the food manufacturer works toward improving the nutritional profile of its business and products. This is achieved by evaluating and reformulating products to deliver on (Patton 2013):

- Reducing calories, fat, saturated fat, trans fat, sugar, or sodium by 10% or more.
- Increasing beneficial nutrients, including vitamins, minerals, and fiber, by 10% or more.
- Formulating products to include at least a half-serving of whole grain, fruit, vegetables, or low or nonfat dairy.

In 2013 alone, General Mills made nutritional improvement to more than 20% of the sales volume in the United States. This was across the businesses of cereals, baking, dairy, meals, and snacks with reductions in sodium and calories and increases in whole grains, among other improvements. Dairy examples include calorie reduction in Yoplait Light from about 100 down to 90 per serving and a new Yoplait Greek 100 yogurt with 100 calories (versus around 130). In the end, this effort adds up to 73% of the entire company's United States sales volume improved since 2005.

These calorie reductions are important progress to addressing overnutrition, especially when the industry comes together with similar efforts. This is what the Healthy Weight Commitment Foundation (HWCF) sets out to do. Fifteen food and beverage companies reduced the calories in the marketplace by 78 per day for every American (Patton 2014). The International Food and Beverage Alliance is a similar group of companies aiming to help companies implement the WHO's global strategy on diet, physical activity, and health. General Mills is involved in both organizations through their commitment to healthier foods.

General Mills also has notable targets to reduce sodium and sugar and increase whole grains. The company is ahead of the pack in bringing sugar down to single-digit grams of sugar per serving for children's cereals, through a 16% decrease (Patton 2013). Sodium reductions are being made across 40% of the product portfolio. The company continues to look ahead toward additional improvements in order to help provide healthy options for consumers.

### 8.1.3 Food security

Food security means there is physical, social, and economic access to sufficient and nutritious food at all times to meets dietary needs for a healthy and active life (Global Food Security Index). Food security has been slowly improving through progress in alleviating poverty and improving market access (Global Food Security Index). However, food security is vulnerable to food price volatility, as evidenced with an all-time peak of hungry and undernourished people during the food price crisis from 2007 to 2009 (FAO 2011). This is because food security is currently not so much about available food but rather an issue of access to affordable food (Wegner & Zwart 2011). Food price volatility is increasing and food prices are projected to be 30 to 50% more over time (International Fund for Agricultural Development [IFAD] 2011; Wegner & Zwart 2011). This brings with it concern for even more of the population being hungry and undernourished (IFAD 2011; Wegner & Zwart 2011). Even if agricultural output increases to meet the estimated need for the population size in 2050 there will still be 300 million hungry due to lack of the means to access the food needed (FAO 2012b).

Currently, there are nearly one billion hungry people in the world (FAO 2012a). Developing countries comprise 98% of the hungry (FAO 2012a). Adding to this, there are around two billion people suffering from "hidden hunger" with micronutrient malnutrition (FAO 2012a). Both types of malnutrition diminish growth and mental development, increase illness, and can lead to early death. In the end, there is a total reduction in earning and learning potential and overall productivity. Hunger and micronutrient deficiencies are severe impediments to social and economic development (FAO 2012a). Just the cost of lost productivity and direct health care costs from malnutrition adds up to 5% of the global gross domestic product (GDP), equivalent to $3.5 trillion per year (FAO 2013a).

Reducing hunger and poverty through building productive agricultural sectors of low-income and highly agriculture-dependent economies is twice as effective as other sector improvements (FAO 2012b). Much of this is because three-quarters of the food insecure population are in rural areas and rely directly or indirectly on agriculture for their livelihoods (Wegner & Zwart 2011). As a result, investment in improved in improved agricultural productivity will be critical to help meet the food needs of a growing population but it also can help bring the rural poor out of poverty to improve their food security. The two key places to focus economic growth and food security are supporting development of small-scale agriculture and women producers (IFAD 2011).

#### 8.1.3.1 Green mountain coffee roasters improving food security

The Vermont-based coffee company Green Mountain Coffee Roasters (GMCR) is building a resilient supply chain with its sustainability program. Working with farmers is the cornerstone of the efforts since that is where the heart of their business lies. In addition to providing a market for responsibly produced coffee, needed financing to support community, and capacity development, the company has dedicated over half of its supply chain projects to fighting widespread hunger at the farm level (GMCR). The company views hunger as a threat to their supply chain that they can help address.

In reviewing their supply chain GMCR found that 67% of coffee growers in Nicaragua, Mexico, and Guatemala suffered from hunger for three to eight months of the year (GMCR). These food-scare times arise between harvest times when there isn't income coming in. GMCR began to partner with non-profit organizations to address this situation. Efforts are focused on increasing home food production and improving storage, diversifying income, and access to markets. Projects have included supporting coffee farmers to grow staple crops and nutritious fruits and vegetables at home or local schools. The company also helped build metal silos for farmers to store and protect grains from humidity and pests and provide year-round access to food. Income diversity from other activities such as eggs or beekeeping have also helped stabilize income throughout the year. Selling surplus produce in the local market has further expanded income potential. This range of solutions came about since GMCR reaches directly to those affected in order to understand what will help them the most. The company believes that this approach has led to their program's success along with collaborating with the right partners such as Mercy Corp (Keller 2012).

GMCR has reached more than 58,000 families across 13 countries since 2007. The deep commitment from GMCR is helping alleviate hunger among its producers and their communities, secure its own coffee supply, and also expand global economic development.

## 8.2 Equity

There is a substantial variance in the ability of different farmers to advance their productivity and improve their income due to access to resources. Developing countries and smallholder producers end up on the end with few resources, both public and private—an inequity. This has led to farmers in developing countries without the same access to breeding improvements, new inputs (fertilizer, chemicals) and management approaches, and other developments (Fraser 2009). To illustrate this, industrialized countries spend 5.16% of agricultural GDP on agricultural research and development for such advancement, whereas developing countries only spend on average 0.56% (Fraser 2009). This research inequity compounds other issues from limited resource access. For example, climate change adaption will be critical for future agricultural production and without research to determine appropriate practices and approaches (e.g., irrigation, breeding, etc.) developing countries will be at a further disadvantage.

The United Nations is developing Sustainable Agriculture Business Principles (SABPs) to provide guidance to the value chain on what businesses can do to help advance toward sustainable agriculture. It is suggested to address equity issues by (Brouwer & Guijt 2013):

- Ensuring economic viability and share value: such as protecting smallholders and eradicating poverty and ensuring market access and fair mechanisms.
- Improving access to and transfer knowledge, skills, and technology: including educating smallholders and investing in local communities.
- Respecting human rights, create decent work, and help rural communities to thrive: such as protecting smallholders and eradicating poverty, investing in local communities, and protecting human rights.
- Encouraging good governance and accountability: such as focus on accountability and anti-corruption and government involvement.

## 8.2.1 Smallholders

Small farms, less than two hectares, account for more than 80% of all farms (Fan 2011). It is estimated that these smallholder farmers produce about 70% of the world's food (Global Compact 2013). The majority of these farmers are in poverty and represent half of the malnourished population (Fan 2011). With little resources available to them it is difficult to improve productivity and get access to markets, credit, and infrastructure to increase their income (Fan 2011). Investment in these farmers can provide multiple benefits including improving global food productivity and overall development. In Vietnam large-scale farms failed to contribute to the development of rural centers since they did not rely on local processing and instead reinvested profits in the cities, however, small-scale farms support the development of rural centers (Fraser 2009). There are also greater yield improvement opportunities and environmental benefits such as reduced soil erosion and better-managed water use with smallholder investment (Fraser 2009).

Smallholders need support. They are often excluded from markets, such as grocery store procurement systems (Fraser 2009). To be included they need access to technology such as irrigation, cold storage and transport, and management (Fraser 2009). This can be done by (WEF 2013):

- Securing small farmers' access and control over productive resources (land, water, forests, seeds, energy).
- Building the capacities of smallholder farmers to:
  - Organize (into associations unions, cooperatives, commodity clusters from local to international levels), network, and engage governments for policy and program work.

    - Increase agricultural productivity through sustainable, agro-ecological production, and post-harvest technologies.
    - Negotiate for and enter into more inclusive business arrangements with other parties (cooperatives, private companies, government, other NGOs) such that these arrangements give farmers more ownership of their business, more influence in decision making over business matters, and a fairer share of risks and rewards.
- Improving physical infrastructures as well as services in rural areas (farm-to-market roads, irrigation, communication system, education, health care, advisory services, insurance, financial services).

### 8.2.2 Women empowerment

Women represent about 50% of the agricultural labor force (Fraser 2009). This ends up equating to approximately two-thirds of the female labor force in developing countries (Fraser 2009). Yet, compared with their male counterparts, female farmers in all regions control less land and livestock, make far less use of improved seed varieties and purchased inputs such as fertilizers, are much less likely to use credit or insurance, have lower education levels, and are less likely to have access to extension services (FAO 2011). Women only receive about 1% of all agricultural financing/credit (Fraser 2009). This limits their ability to invest in productivity improvement such as purchasing irrigation or new seed varieties.

Women would achieve the same yields as men if they had equal access to agricultural resources and services (FAO 2011). However, due to their limited access, women average a 25% lower yield compared to men (FAO 2011). If this gap were filled an increase in agricultural output of 2.5 to 4% would result with the land women currently control (FAO 2011). This then could reduce the number of undernourished people by 12 to 17% (FAO 2011). The benefits reach beyond this, however. Improved income for women brings better child nutrition, health, and education such that the FAO affirms that empowering women is a well-proven strategy for improving children's well-being (FAO 2011).

The following are key areas for investment to lead to these benefits including improving women's (Chan 2010; FAO 2011; & Fraser 2009):

- Land and livestock ownership
- Education and training (literacy and agricultural information)
- Financial services
- Sourcing contracts
- Technology
- Fair labor

### *8.2.2.1 Starbucks investing in agricultural communities*

Starbucks Corporation reaches more than 60 countries with 19,000 retail coffee shops (Starbucks 2014). Notably, the company's global reach has been paralleled by its commitment to sustainability. At the heart of the program is a sustainable sourcing program that reaches to the farm level to help build capacity and community development. These efforts are centered around five main initiatives: ethical sourcing, farmer loans, farmer support centers, social development investments, and collaborative relationships.

Starbucks has a goal to deliver $20 million of loans to farmers by 2015. These loans are designed to fill gaps left from traditional financing sources. Part of the funding provides smallholders financing and technical assistance through the Fairtrade Access Fund for Fairtrade producer organizations and cooperatives. This has helped farmers and cooperatives gain improved access to international markets, a minimum price safety net, and other economic benefits such as a cash flow buffer during harvest time when they need to pay their farmers and workers (Starbucks).

Starbucks established six Farmer Support Centers in key coffee-growing regions. Each center is staffed with experts that can provide assistance directly to farmers. The center in Costa Rica includes an innovative research and development program on a 240-hectare farm to evaluate and identify best management practices and improved coffee varietals (Starbucks 2013). The insights gained are then shared across each center and Starbucks' producers.

Starbucks has a history of working collaboratively to advance sustainability solutions. In 2013, Starbucks partnered with United States Agency for International Development (USAID) to expand the technical assistance provided at the Columbian Farmer Support Center in Manizales. The partnership extended training to an additional 25,000 farmers in Columbia's most vulnerable regions (USAID 2013). Starbucks also works closely with organizations to implement social development projects. For example, collaborating with Save the Children helped bring education to children and resources to schools in Guatemala (Starbucks b).

All of these efforts are underpinned by Starbucks' commitment to purchase only ethically sourced coffee by 2015. Importantly, the company provides a market for coffee produced by the farmers and communities supported through their other efforts. This is executed through the Coffee and Farmer Equity (CAFE) Practices. CAFE includes comprehensive buying guidelines that ensure quality while also delivering on social, economic, and environmental standards such as (Starbucks c):

- Product quality. All coffee must meet standards for high quality.
- Economic accountability. Economic transparency is required. Suppliers must submit evidence of payments made throughout the coffee supply chain to demonstrate how much of the price paid for green coffee gets to the farmer.

- Social responsibility. Measures evaluated by third-party verifiers help protect the rights of workers and ensure safe, fair, and humane working and living conditions. Compliance with minimum-wage requirements and prohibition of child and forced labor is mandatory.
- Environmental leadership. Measures evaluated by third-party verifiers help manage waste, protect water quality, conserve water and energy, preserve biodiversity, and reduce agrochemical use.

The program was developed in Collaboration with Conservation International and has had notable success. For example, farmers employing CAFE practices maintain or improve soil fertility and all school-age children on smallholder farms to attend school. The cocoa program requires that women and men farmers get paid equally (Starbucks 2009). Ninety-three percent of the 545 million pounds of coffee purchased by Starbucks in 2012 was ethically sourced (Starbucks c). Starbucks' comprehensive approach to supporting farmer communities is a leading example in the market that delivers widespread benefits that help build the business.

### 8.2.3 Farm labor

Agricultural production relies on the hands of workers in the field. Wages for this work are low relative to other sectors and conditions are often hazardous (Fraser 2009). In the United States the average waged worker income was just $10,000 to $12,499 (McCluskey, McGarity, & Shapiro 2013). The International Labour Organization of the United Nations (ILO) estimates that there are 170,000 agricultural worker deaths on the job each year (ILO 2009). Farm worker fatality rates in the United States are seven times higher than the private industry average and injury rates are 20% higher than average (McCluskey et al. 2013). Hazards come from the physically demanding work and use of dangerous equipment. Farmers are also exposed to potentially harmful pesticides in greater quantity, concentration, and frequency than the rest of the population. Yet, laws and enforcement are extremely limited to support these workers. Farm workers are not covered under the United States' labor legislation (the National Labor Relations Act), to limit protection to the right to organize and bargain collectively, overtime, child labor, unemployment insurance, and worker compensation (Fraser 2009). Labor expenses comprise about 17% of total variable farm costs and as much as 40% of costs for labor-intensive crops such as fruit and vegetables (United States Department of Agriculture Economic Research Service [USDA ERS]). With labor representing a large portion of a farm's costs there is preference in utilizing short-term and contract workers. This reduces the accountability the employer has to its workers to provide proper training, tools, and other practices to address dangers (McCluskey et al. 2013).

To compound this situation, farm labor is not well-equipped to advocate for better conditions. The average agricultural worker stopped school at seventh grade and cannot speak English well (McCluskey et al. 2013). About half of farm workers in the United States are undocumented immigrants (Kelly & Lang 2012). Only 2% of farm workers belong to unions that could aid in communicating worker needs (McCluskey et al. 2013).

In some areas, such as the United States, farm labor shortages are increasing. This can drive up wages and bring other benefits for workers (Wozniacka 2013). However, there is an alternate, unethical path that unfortunately is a reality—forced labor and slavery in agriculture remains a practice in parts of the world. Cacao production, for example, has widespread child slavery and abusive labor practices (Gregory 2013). Developing countries are not the only sources of such atrocities. In 2010 more than 400 Thai farmers were trafficked into the United States to forcibly work on farms and orchards (Bon Appetit Management Company Foundation and United Farm Workers 2011).

Enforcing ethical, safe, and fair labor in the supply chain is critical. Product manufacturers have adopted code of conduct programs to begin to meet this need. Companies include legal, safety, and labor requirements in the codes and require suppliers to meet them. This is best enforced with site audits. Since the needs for code of conduct programs are very similar across companies, the Supplier Ethical Data Exchange (Sedex) provides a way for companies to share audit results. The Ethical Trading Initiative is another organization that encourages alignment on code of conduct programs. Code of conduct efforts have potential to have a positive impact on worker health and safety, working hours, wages, and the use of child labor (Fraser 2009). However, typically codes of conduct are less effective at addressing the freedom of association, discrimination, regular employment, and harsh treatment (Fraser 2009). The standards should be sure to address the broad range of labor issues. Companies that enforce their code of conduct and supplement it with supplier support and training end up with greater success and business benefit.

Table 8.1 outlines criteria from Oxfam International for guaranteeing the right to decent work and respect for the human rights of agricultural laborers based on a number of ILO conventions and typical laws. The FAO Sustainability Assessment of Food and Agriculture systems (SAFA) provides tools including indicators and actions to review and integrate these considerations into supply chain programs (FAO 2013b). However, this issue is not isolated to farms, processing facilities need to ensure the same rights to workers (Barclay 2013).

#### *8.2.3.1 Bon Appétit partnership with the coalition of immokalee workers*

Bon Appétit Management Company (BAMC a) runs 550 on-site restaurants at various institutions across the United States. BAMC has been on the forefront of sustainability through its history including a local food program in

**Table 8.1** Criteria for guaranteeing the right to decent work and response for the human rights of agricultural laborers

| |
|---|
| Compliance with national laws and ratified international laws—whichever are better on employment conditions and workers' rights. |
| Employees are provided with legal contracts. |
| Employees are informed about their rights and there is a mutually agreed and documented system for dealing with complaints and grievances. |
| Subcontractors meet the same criteria. |
| Freedom of association and right to collective bargaining (as established in ILO conventions) are guaranteed. Where this is restricted by law, employers should facilitate alternative independent means of free association and collective bargaining. |
| No child labor (as established in ILO conventions). |
| Health and Safety rules are applied (according to ILO convention 155 on Occupational Health and Safety). |
| Fair wages and compensation rules: |
| Workers must be paid wages at least equivalent to the legal national minimum wage or the relevant industry standard, whichever is higher. In any event, wages should always be enough to meet basic needs and to provide some discretionary income. In instances of piecework, the pay rate must permit the worker to earn at least the minimum wage or relevant industry standard (whichever is higher) during normal working hours and under normal working conditions. |
| Working hours are not excessive: they comply with national laws, and benchmark industry standards, whichever affords greater protection. In any event, workers shall not on a regular basis be required to work in excess of 10 hours per day. Overtime shall be voluntary, shall not be demanded on a regular basis, and shall always be compensated at a premium rate. |
| No discrimination on the basis of: race, caste, nationality, religion, disability, gender, sexual orientation, union membership, political affiliation, age, marital status, working status (i.e., temporary, migrant, seasonal), HIV/AIDS (in accordance with ILO conventions). |
| No forced labor (as defined in ILO conventions). |

The material is adapted by the publisher from Harnessing Agriculture for Development, Figure 7: Criteria for Guaranteeing the Right to Decent Work and Respect for the Human Rights of Agricultural Labourers [http://www.oxfam.org/sites/www.oxfam.org/files/bp-harnessing-agriculture-250909.pdf] with the permission of Oxfam GB, Oxfam House, John Smith Drive, Cowley, Oxford OX4 2JY UK www.oxfam.org.uk Oxfam GB does not necessarily endorse any text or activities that accompany the materials, nor has it approved the adapted text, Fraser 2009.

1999 that required 20% of ingredients to be sourced from small farmers within 150 miles (BAMC a). BAMC added to its legacy by taking on farm labor abuses and becoming the first food service company to partner with the Coalition of Immokalee Workers (CIW) to establish and enforce a strict code of conduct for tomato growers in 2009.

The company decided to take action after learning that United States civil rights officials had prosecuted seven slavery operations involving over 1,000 workers in Florida's fields since 1997 (BAMC b). In 2008, six people were convicted of imprisoning more than a dozen men who were chained, beaten, and forced to work (Black 2009).

At the front line of these battles was the CIW. The CIW is a community-based farmworker organization in Immokalee, Florida, on a mission to advance farm worker conditions to modern-day standards with fair treatment in accordance with national and international labor standards. BAMC developed an agreement with CIW to enforce more humane working conditions and treatment for the farm workers that includes (BAMC b):

- A "Minimum Fair Wage." Workers are paid a wage premium that reflects the unique rigors and uncertainty of farm labor.
- An end to traditional forms of wage abuse. Standards require growers to implement time clocks and to reconcile wages paid with pounds harvested, workers are paid for every hour worked and every pound picked.
- Worker empowerment. Workers are informed of their rights through a system jointly developed by the growers and the CIW. Growers also collaborate with the CIW and BAMC to implement and enforce a process for workers to pursue complaints without fear of retribution.
- Worker safety. A worker-controlled health and safety committee give farmworkers a voice in addressing potentially dangerous working conditions, including pesticide, heat, and machinery issues.
- Third-party monitoring. Growers must permit third-party monitoring that includes worker participation.

A ground-breaking component of this effort was that the farm workers get a premium paid by BAMC immediately and directly (Black 2009). This can add up to a notable raise for many workers. This model has had great success and expanded to be a separate program called the Fair Food Program (FFP). The FFP continues to provide a comprehensive, verifiable, and sustainable approach to ensure better wages and working conditions in Florida's tomato fields. Ninety percent of Florida's tomato farms have implemented the FFP (CIW). Food service and retail companies participate in the FFP paying premiums for this labor assurance including, beginning in 2014, Walmart.

BAMC continues to push progress on labor issues through consumer education and empowering farm workers. As a member of the Equitable Food Exchange, the company is expanding its efforts to more geographic regions and crops.

## 8.2.4 Local food

The concept of local food refers to shorter distances traveled generally within a community, state, or region. One hundred or 300 miles is often used but does not represent the entire concept of local foods since it more generally represents

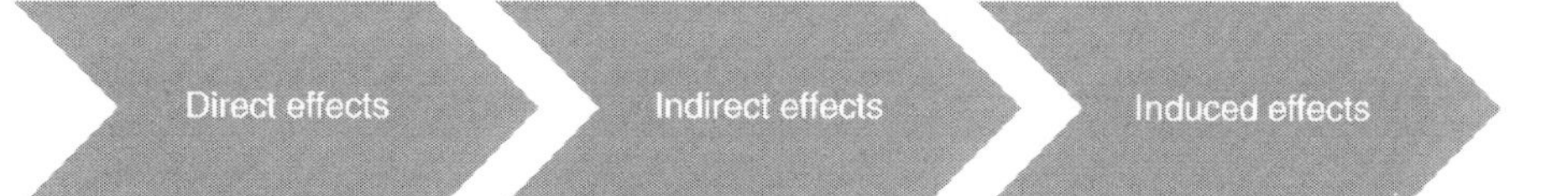


**Direct effects**
- Value of new production, processing, and retail output
- Additional jobs and labor income generated

**Indirect effects**
- Value of locally supplied inputs and services by businesses that service the producers
- Additional jobs and labor income generated

**Induced effects**
- Value of employee and business spending in local economy
- Additional jobs and labor income generated

**Figure 8.2** Range of benefits from local food systems (USDA ERS 2010).

food that has a more direct connection to the consumer in part because of the distance traveled. Local food is often used as a proxy for less impact from food transportation. However, this is not an accurate view since food transport is a minor impact in the food system and the distance traveled is not necessarily connected to a better environmental profile, as discussed in a previous chapter. The concept of supporting community and regional food systems, however, provides important social and economic benefits such as increased employment and income (see Figure 8.2) (USDA ERS 2010).

In the United States, there has been a decrease in the number of farmers and farms (farms are now larger) reducing the levels of employment, local retail spending, and local per capita income in rural communities (Copeland & Zin, 1998). All of the apples consumed in Iowa in 1970 were produced in the state, but by 1999 Iowa farmers grew only 15% of the apples consumed in the state (Pirog et al. 2001). This type of shift limits the potential for local self-sufficiency and increases dependency on outside sources, whereas local food systems develop the community with employment and income in a multiplier effect (see Figure 8.2) (USDA ERS 2010). Local food also can have a positive impact on their communities, including the following (FAO 2012a):

- Greater availability of more nutritious and more diverse local fruits and vegetables.
- Reduction of food safety risks as a result of decentralized production.
- Conservation of farmland.
- Provision of incentives for small farmers to stay in rural areas, instead of moving to the cities without employment.
- Development of a sense of pride and social cohesion in a community.
- Protection of agricultural biodiversity.

### *8.2.4.1 Locally grown produce at Chipotle Mexican Grill*

Chipotle Mexican Grill has more than 900 quick-service restaurants throughout the United States with a focus on "food with integrity." The limited service menu is prepared from fresh and sustainable ingredients. The company began integrating locally sourced products into the business in 2007 (Wederquist). Along the way, the company also hopes to support the development of family farms and the local economy.

Chipotle has a high standard for ingredient quality and believes that local foods provide fresher, better tasting food. The company looks to source more ingredients for its menu from at least 350 miles away, what they consider local. What is sourced for each restaurant depends on what is available in its own region. In some areas local meat or dairy is available but the company is focused on at least having every store include local produce on its menu.

Chipotle's local produce program includes red onions, oregano, bell peppers and jalapeno peppers, and romaine (Wederquist). Other produce for the menu is typically harder to get across the country due to the climatic requirements for things such as avocados and limes. The company sources from farmers or farmer cooperatives with the size/scale needed to supply at least a portion of restaurants in one its markets while in season. This typically means the grower delivers, via refrigerated truck, to one of the company's 23 produce houses where produce shipments are consolidated, checked for quality, and then delivered to distribution centers in each area to then head to the restaurants (Wederquist).

Chipotle requires its local producers to have effective food safety programs in place. This is verified through a third-party audit that is compliant with the standardized good agricultural practices (GAP)—such as a USDA or private audit (Wederquist). Chipotle then follows with its own site visit before bringing on a new producer. The company usually works with mid-sized family farms, rather than small farms, but a range of farms are included in the program (Wederquist).

The company has faced challenges with scaling beyond produce because of lack of infrastructure—there aren't enough slaughter facilities to source local chicken in all markets, for example. Chipotle has also found a limit to the amount of produce it can source locally due to quality issues from weather/pests and the seasonality of produce.

Already the company sources over 90% of the produce in the program from within 250 miles of distribution centers, 70% from within 150 miles, and one-third from less than 50 miles (Wederquist). This added up to five million pounds of locally sourced produce in 2010. The restaurants promote the locally sourced items since Chipotle sources directly from the producer and can trace it all the way to the restaurant. For example, Pahl's fifth-generation family farm in Apple Valley, Minnesota, provides 85% of the bell pepper needs for about 50 restaurants in the area (Wederquist).

Chipotle started their local sourcing program out small and has expanded it each year. The company continues to grow the program and also seeks to reduce the total distance between the farm and the restaurant to further support freshness and local economy development. The company strongly holds the belief that the more their customer sees the benefits of eating food from sustainable sources, the more they will want it from everyone and it will need to become the norm.

## 8.3 Summary

Nutrition, food security, and producer equity are closely connected. Efforts to improve malnutrition across the supply chain (see Table 8.2) cross over in addressing the other issues. Investing in small-scale businesses and women in agriculture is critical to addressing food security and long-term supply needs and is one of the highest-yielding investment opportunities to improve food availability. Effective investment includes developing the capacity of producers in a region, improving access to agricultural resources, and developing skills and needed infrastructure. Assisting smallholders and women producers to meet market standards is then best followed with sourcing agreements to provide market access. However, fundamental human rights issues cannot be neglected across farms of all sizes and through the supply chain. In the end, we can build a system that supports the capacity of its workers, enhances economic development, and feeds the population with wholesome foods (Tables 8.3, 8.4, and 8.5).

**Table 8.2** Improvement opportunities to address malnutrition (of all forms) (FAO 2013a)

| | |
|---|---|
| **Agricultural production** | • Crop and livestock diversity<br>• Food safety<br>• Reduce food waste<br>• Smallholder and women support and empowerment |
| **Manufacturing** | • Food safety<br>• Product formulation for better nutritional profile (e.g., additives, fortification, processing)<br>• Reduce food waste |
| **Consumer** | • Consumer education and labeling<br>• Appropriate advertising and marketing (especially to children)<br>• Targeted food assistance<br>• Reduce food waste |

**Table 8.3** Summary of the principles–practices–potential of sustainability for food safety and nutrition

| | |
|---|---|
| **Principle** | *Safe and highly nutritious food is accessible and affordable to promote and support a healthy population* |
| **Practices** | • Avoid microbial, chemical, and allergen contamination<br>• Reduce fat, sugar, and salt in processed foods<br>• Provide understandable and meaningful information about food to consumers<br>• Market food responsibly, especially to children and teens<br>• Ensure diverse, healthy, and nutritious food is available and affordable |
| **Potential** | • Wholesome food supply |

**Table 8.4** Summary of the principles–practices–potential of sustainability for producer equity

| | |
|---|---|
| **Principle** | *Producer equity and rural economy and development are strengthened with fair and responsible production and sourcing.* |
| **Practices** | • Support smallholders and women in agriculture:<br>   • Ensure economic viability and share value<br>   • Build capacity<br>   • Improve infrastructure and rural services<br>• Source from local communities<br>• Integrate diversity of sources into procurement (with longer-term commitments for smallholders and women when possible) |
| **Potential** | • Enhanced economic development<br>• Resilient food supply<br>• Food security |

**Table 8.5** Summary of the principles–practices–potential of sustainability for workers

| | |
|---|---|
| **Principle** | *Safe and suitable working conditions are provided to support employees across the supply chain* |
| **Practices** | • Respect human rights<br>• Provide fair wages and labor conditions<br>• Empower workers<br>• Provide safe working conditions |
| **Potential** | • Build capacity<br>• Deliver safe and prosperous livelihoods |

## Resources

Global Food Safety Initiative http://www.mygfsi.com
Oxfam International: www.oxfam.org
Supplier Ethical Data Exchange: http://www.sedexglobal.com/

## References

Barclay, E. (2013). *Why slave labor still plagues the global food system*. Retrieved from http://www.npr.org/blogs/thesalt/2013/06/19/193548623/why-slave-labor-still-plagues-the-global-food-system

Black, J. (2009). *Putting the squeeze on tomato growers to improve conditions for farm workers.* Retrieved from http://www.washingtonpost.com/wp-dyn/content/article/2009/04/28/AR2009042800835.html

Bon Appetit Management Company (BAMC a). *History*. Retrieved from http://www.bamco.com/timeline/

Bon Appetit Management Company (BAMC b). *CIW fair food agreement*. Retrieved from http://www.bamco.com/timeline/ciw-fair-food-agreement/

Bon Appetit Management Company Foundation and United Farm Workers. (2011). *Inventory of farmworker issues and protections in the United States*. Retrieved from http://www.oxfamamerica.org/static/oa3/files/inventory-of-farmworker-issues-and-protections-in-the-usa.pdf

Brouwer, H., & Guijt, J. (2013). *Moving towards sustainable agriculture business principles.* Retrieved from http://www.unglobalcompact.org/docs/issues_doc/agriculture_and_food/SABP_1st_Consultation_Synthesis_Paper.pdf

CBC. (2013). *Food waste, overeating threaten global security*. Retrieved from http://www.huffingtonpost.ca/2013/11/23/food-waste-overeating_n_4328424.html?utm_hp_ref=fb&src=sp&comm_ref=false

Chipotle Mexican Grill. *Environment*. Retrieved from http://www.chipotle.com/en-us/fwi/environment/environment.aspx

Coalition of Immokalee Workers (CIW). *Fair Food Program*. Retrieved from http://ciw-online.org/fair-food-program/).

Consultative Group on International Agricultural Research (CGIAR). (2014). *Undernourishment and obesity*. Retrieved from http://ccafs.cgiar.org/bigfacts2014/#theme=food-security&subtheme=undernourishment

Chan, M-K. (2010). *Improving opportunities for women in smallholder-based supply chains: Business case and practical guidance for international food companies*. Retrieved from http://www.gatesfoundation.org/learning/Documents/gender-value-chain-guide.pdf

The Executive Office of the President of the United States (2010) Strengthening the Rural Economy - The Current State of Rural America II. THE CURRENT STATE OF RURAL AMERICA http://www.whitehouse.gov/administration/eop/cea/factsheets-reports/strengthening-the-rural-economy/the-current-state-of-rural-america

Fan, S. (2011). *Leveraging smallholder agriculture for development*. Retrieved from http://www.slideshare.net/shenggenfan/leveraging-smallholder-agriculture-for-development

Food Allergy Research and Education (FARE). *Food allergy facts and statistics for the U.S.*. Retrieved from http://www.foodallergy.org/document.doc?id=194

Food and Agriculture Organization of the United Nations (FAO). *AGP—highly hazardous pesticides (HHPs).* Retrieved from http://www.fao.org/agriculture/crops/thematic-sitemap/theme/pests/code/hhp/en/

Food and Agriculture Organization of the United Nations (FAO). (2011). *The state of food and agriculture*. Retrieved from http://www.fao.org/docrep/013/i2050e/i2050e.pdf

Food and Agriculture Organization of the United Nations (FAO). (2012a). *Improving food systems for sustainable diets in a green economy*. Retrieved from http://www.fao.org/fileadmin/templates/ags/docs/SFCP/WorkingPaper4.pdf

Food and Agriculture Organization of the United Nations (FAO). (2012b). *Towards the future we want: End hunger and make the transition to sustainable agricultural and food systems*. Retrieved from http://www.fao.org/docrep/015/an894e/an894e00.pdf

Food and Agriculture Organization of the United Nations (FAO). (2013a). *The state of food and agriculture.* Retrieved from http://www.fao.org/docrep/018/i3300e/i3300e.pdf

Food and Agriculture Organization of the United Nations (FAO). (2013b). *Sustainability assessment of food and agriculture systems: Indicators*. Retrieved from http://www.fao.org/fileadmin/templates/nr/sustainability_pathways/docs/SAFA_Indicators_final_19122013.pdf

Forini, K., Denison, R., Stiffler, T., Fitzgerald, T., & Goldburg, R. (2005). *Resistant bugs and antibiotic drugs: State and county estimates of antibiotics in agricultural feed and animal waste*. Retrieved from www.edf.org/health/report/resistant-bugs-and-anitbiotic-drugs

Fraser, A. (2009). *Harnessing agriculture for development*. Retrieved from http://www.oxfam.org/sites/www.oxfam.org/files/bp-harnessing-agriculture-250909.pdf

The Global Compact. (2013). UN *Global Compact sustainable agriculture business principles: White paper.* Retrieved from http://www.unglobalcompact.org/docs/issues_doc/agriculture_and_food/SABP_White_Paper_July13.pdf

Global Food Security Index. *Our methodology and expert panel*. Retrieved from http://foodsecurityindex.eiu.com/Home/Methodology

Green Mountain Coffee Roasters (GMCR). *Working with farmers*. Retrieved from http://www.gmcr.com/Sustainability/ResilientSupplyChain/WorkingwithFarmers/WorkingwithFarmers.aspx

Gregory, A. (2013). *Chocolate and child slavery: Say no to human trafficking this holiday season.* Retrieved from http://www.huffingtonpost.com/amanda-gregory/chocolate-and-child-slave_b_4181089.html

Horrigan, L., Lawrence, R., & Walker, P. (2002). How sustainable agriculture can address the environmental and human health harms of industrial agriculture. *Environmental Health Perspectives* 110(5), 445–456.

International Agency for Research on Cancer of the World Health Organization (IARC). (2014). *IARC monographs on the evaluation of carcinogenic risks to humans.* Retrieved from http://monographs.iarc.fr/ENG/Classification/

International Fund for Agricultural Development (IFAD). (2011). *IFAD strategic framework 2011–2015.* Retrieved from http://www.ifad.org/sf/strategic_e.pdf

International Labour Organization (ILO). (2009). *Agriculture: A hazardous work*. Retrieved from http://www.ilo.org/safework/areasofwork/hazardous-work/WCMS_110188/lang--en/index.htm

Jenkin, G., Madhvani, N., Signal, L., & Bowers, S. (2014). A systematic review of persuasive marketing techniques to promote food to children on television. *Obesity Reviews.* doi: 10.1111/obr.12141

Keller, C. (2012). *Green Mountain Coffee Roasters case study: Fighting hunger, growing communities.* Retrieved from http://ccc.uschamber.com/blog/2012-10-16/green-mountain-coffee-roasters-case-study-fighting-hunger-growing-communities

Kelly, M., & Lang, H. (2012). *Worker equity in food and agriculture: Executive summary.* Retrieved from http://www.sustainalytics.com/sites/default/files/workerequity_october2012.pdf

Mathew, A., Cissell, R., & Liamthong, S. (2007). Antibiotic resistance in bacteria associated with food animals: A United States perspective of livestock production. *Foodborne Pathogens and Disease* 4(2),115–133.

McCluskey, M., McGarity, T., & Shapiro, S. (2013). *At the company's mercy: protecting contingent workers from unsafe working conditions*. Retrieved from http://www.progressivereform.org/articles/Contingent_Workers_1301.pdf

Patton, K. (2013). *Health improvements across all U.S. categories*. Retrieved from http://www.blog.generalmills.com/2013/10/health-improvements-across-all-u-s-categories/)

Patton, K. (2014). *6.4 trillion calorie reduction*. Retrieved from http://www.blog.generalmills.com/2014/01/6-4-trillion-calorie-reduction

Pirog, R., Van Pelt, T., Enshayan, K., Cook, E. (2001) Food, fuel, and freeways: an Iowa perspective on how far food travels, fuel usage, and greenhouse gas emissions. http://www.leopold.iastate.edu/sites/default/files/pubs-and-papers/2011-06-food-fuel-and-freeways-iowa-perspective-how-far-food-travels-fuel-usage-and-greenhouse-gas-emissions.pdf

Public Health Law Center (PHLC). *Food marketing to kids*. Retrieved from http://publichealthlawcenter.org/topics/healthy-eating/food-marketing-kids

Schab D. W., & Trinh, N. H. (2004). Do artificial food colors promote hyperactivity in children with hyperactive syndromes? A meta-analysis of double-blind placebo-controlled trials. *Journal of Developmental and Behavioural Pediatrics* 25, 423–434.

Shelke, K., Van Wart, J., & Francis, C. (2009). Social aspects of the food supply. In C. Baldwin (Ed.), *Sustainability in the food industry* (pp.145-158). Ames, IA: Wiley-Blackwell.

Starbucks Corporation a. *Farmer loans programs*. Retrieved from http://www.starbucks.com/responsibility/community/farmer-support/farmer-loan-programs

Starbucks Corporation b. *Social development investments*. Retrieved from http://www.starbucks.com/responsibility/community/farmer-support/social-development-investments

Starbucks Corporation c. *Coffee*. Retrieved from http://www.starbucks.com/responsibility/sourcing/coffee

Starbucks Corporation. (2009). *Cocoa practices: Evaluation guidelines version 1.3*. Retrieved from http://www.starbucks.com/assets/4586dcbfda0d46738fca5211c017a8c3.pdf

Starbucks Corporation. (2013). *Starbucks expands $70 million ethical sourcing program with new global agronomy center*. Retrieved from http://www.businesswire.com/news/home/20130318006738/en/Starbucks-Expands-70-Million-Ethical-Sourcing-Program#.UwFwMfb-NNE

Starbucks Corporation. (2014). *Starbucks company profile.* Retrieved from http://globalassets.starbucks.com/assets/e12a69d0d51e45d58567ea9fc433ca1f.pdf

United Nations Environment Programme (UNEP). (2012). *The critical role of global rood consumption patterns in achieving sustainable food systems and food for all.* Retrieved from http://fletcher.tufts.edu/~/media/Fletcher/Microsites/CIERP/Publications/2012/UNEP%20Global%20Food%20Consumption.pdf

U.S. Agency for International Development (USAID). (2013). *A new partnership to support Colombia's coffee farmers*. Retrieved from http://blog.usaid.gov/2013/08/new-partnership-to-support-colombias-coffee-farmers/

U.S. Centers for Disease Control and Prevention (CDC). (2012). *Foodborne illness, foodborne disease (sometimes called "food poisoning").* Retrieved from http://www.cdc.gov/foodsafety/facts.html

U.S. Centers for Disease Control and Prevention (CDC). (2013a). *Infections from some foodborne germs increased, while others remained unchanged in 2012*. Retrieved from http://www.cdc.gov/media/releases/2013/p0418-foodborne-germs.html

U.S. Centers for Disease Control and Prevention (CDC). (2013b). *Antibiotic resistance threats in the U.S.* Retrieved from http://www.cdc.gov/features/antibioticresistancethreats/

U.S. Centers for Disease Control and Prevention (CDC). (2013c). *Adult obesity facts.* Retrieved from http://www.cdc.gov/obesity/data/adult.html

U.S. Centers for Disease Control and Prevention (CDC). (2013d). *A growing problem.* Retrieved from http://www.cdc.gov/obesity/childhood/problem.html

U.S. Department of Agriculture Economic Research Service (USDA ERS). *Farm labor.* Retrieved from http://www.ers.usda.gov/topics/farm-economy/farm-labor.aspx#.Uv_c-vb-OC8

U.S. Department of Agriculture Economic Research Service (USDA ERS). (2010). *Local food systems: Concepts, impacts, and issues.* Retrieved from http://www.ers.usda.gov/publications/err-economic-research-report/err97.aspx#.UwQuTvb-NNG

U.S. Food and Drug Administration (FDA). (2013a). *FDA annual report on antimicrobials sold or distributed for food-producing animals in 2011.* Retrieved from http://www.fda.gov/AnimalVeterinary/NewsEvents/CVMUpdates/ucm338178.htm

U.S. Food and Drug Administration (FDA). (2013b). *FDA targets trans fat in processed foods.* Retrieved from http://www.fda.gov/ForConsumers/ConsumerUpdates/ucm372915.htm

U.S. Food and Drug Administration (FDA). (2009). *Food allergies: Reducing the risks.* Retrieved from http://www.fda.gov/ForConsumers/ConsumerUpdates/ucm089307.htm

University of California Los Angeles Food and Drug Allergy Care Center. *About allergies: Why are allergies increasing.* Retrieved from http://fooddrugallergy.ucla.edu/body.cfm?id=40

Wederquist, H. *Towards local and regional sourcing: Sysco and Chipotle webinar.* Retrieved from http://ngfn.org/resources/ngfn-cluster-calls/towards-local-and-regional-sourcing-sysco-and-chipotle

Wegner, L, & Zwart, G. (2011). *Who will feed the world: The production challenge.* Retrieved from http://www.oxfamnovib.nl/Redactie/Downloads/Rapporten/who-will-feed-the-world-rr-260411-en.pdf

Whole Foods Market (WFM). (2013). *Whole Foods Market announces enhanced standards for fresh produce and flowers.* Retrieved from http://media.wholefoodsmarket.com/news/produce-rating-release

World Economic Forum (WEF). *The dangers of hubris on human health.* Retrieved from http://reports.weforum.org/global-risks-2013/risk-case-1/the-dangers-of-hubris-on-human-health/

World Economic Forum (WEF). (2013). *Invest in smallholder farmers.* Retrieved from http://forumblog.org/2013/06/invest-in-smallholder-farmers/)

World Health Organization (WHO). *Global strategy on diet, physical activity and health.* Retrieved from http://www.who.int/dietphysicalactivity/strategy/eb11344/strategy_english_web.pdf

World Health Organization (WHO). (2009). *The WHO recommended classification of pesticides by hazard and guidelines to classification 2009.* Retrieved from http://www.who.int/ipcs/publications/pesticides_hazard/en/)

World Health Organization (WHO). (2012). *The evolving threat of antimicrobial resistance.* Retrieved from http://www.who.int/patientsafety/implementation/amr/publication/en/index.html

World Health Organization (WHO). (2013). *Obesity and overweight.* Retrieved from http://www.who.int/mediacentre/factsheets/fs311/en/

Wozniacka, G. (2013). *Farmers face labor shortages as workers find other jobs.* Retrieved from http://www.huffingtonpost.com/2013/09/26/farm-labor-shortages_n_3996502.html

Yale Rudd Center for Food Policy and Obesity (Rudd). (2012). *Trends in television food advertising to young people.* Retrieved from http://www.yaleruddcenter.org/resources/upload/docs/what/reports/RuddReport_TVFoodAdvertising_5.12.pdf

# 9 Sustainable Food Consumption and the Potential of the Principles

*Principle: The supply chain and consumers advance sustainable business and food consumption.*

Consumer diet and behavior are critical contributors to the sustainability of the food system. These decisions are partly in the hands of the population but are influenced by upstream players such as food manufacturers. In the end, each stage in the supply chain plays a role in advancing sustainability, from farmer to consumer. The Ten Principles discussed in this book identify the key sustainability needs across the supply chain, and they offer exciting potential to shift from a system that depletes to one that improves the economy, environment, and society.

## 9.1 Sustainable consumption

Consumer interest in sustainability continues to grow. In the United States, 70% of consumers are seeking out more sustainable products (Shelton Group 2012). However, consumers generally do not know what this means to them or their purchases. The supply chain can play an important role in helping consumers understand their impacts and making it easier for them to make the best choice. This is complicated since there is not a single solution. For example, diets that are based more on locally produced, seasonal foods that are grown using energy-efficient management systems require little cooking time and diets that include relatively low amounts of meat and dairy products would result in overall reductions in energy demands (Schneider & Smith 2009). Consumer education is necessary, from what to choose to eat to how to prepare it. It is also important to connect the consumer to the farm so they can better understand the value of sustainable choices. Engaging the consumer and

*The 10 Principles of Food Industry Sustainability*, First Edition. Cheryl J. Baldwin.

**Table 9.1** Summary of the principles–practices–potential of sustainability for consumers and the supply chain

| | |
|---|---|
| **Principle** | *The supply chain and consumers advance sustainable business and food consumption.* |
| **Practices** | • Optimize consumer's:<br>  • Eating<br>    • Shift diets to more plant-based and fresh foods, and moderate animal-based and highly-processed food intake<br>    • Eat a balanced diet with a variety of foods to maintain a heathy body weight<br>    • Purchase responsibly-produced foods<br>  • Transportation<br>  • Storing and preparing<br>  • Food waste minimization |
| **Potential** | • Sustainable food consumption<br>• Support each of the other areas of potential |

shifting them to more sustainable diets and behaviors is essential to improving the sustainability of the food supply chain (see Table 9.1).

## 9.1.1 Sustainable diets and behaviors

One of the most critical sustainability decisions made by a consumer is what they purchase to eat. The current choices consumers are making need to align better with sustainability since they are generally headed in the wrong direction. Diets are moving to more processed and packaged foods at the same time as meat and dairy product consumption is increasing (Canning, Charles, Huang, Polenske, & Waters 2010; Haley). These trends have already led to increased energy demands in the food supply (Canning et al. 2010). This is compounded by consumption of more animal-based foods having notable effects on overall environmental demands from the food system, especially with red meat being more greenhouse gas (GHG)-intensive than other foods (DeWeerdt). As a result, the Food and Agriculture Organization of the United Nations (FAO) recommends consuming less meat and dairy products and selecting foods that are less processed such as fruit and vegetables or fresh fish (FAO 2012).

Another consideration is the amount of food consumed. Industrialized countries consume more calories than needed, unnecessarily putting extra demand on the food system (CBC 2013). Eating more than needed is a waste of resources since the food is not effectively utilized (Department for Environment, Food, and Rural Affairs for the United Kingdom [DEFRA] 2013). Thus, eating the right amount of food is another strategy to achieve sustainable consumption. Reducing the number of overweight people by half would help close the gap in food needed for the projected population size in 2050 by 6%, due to more food being available for other people to eat (WRI).

Consumers need to waste less food. Reducing wasted food by half would close the gap in food needed to feed the growing population by 20% (WRI).

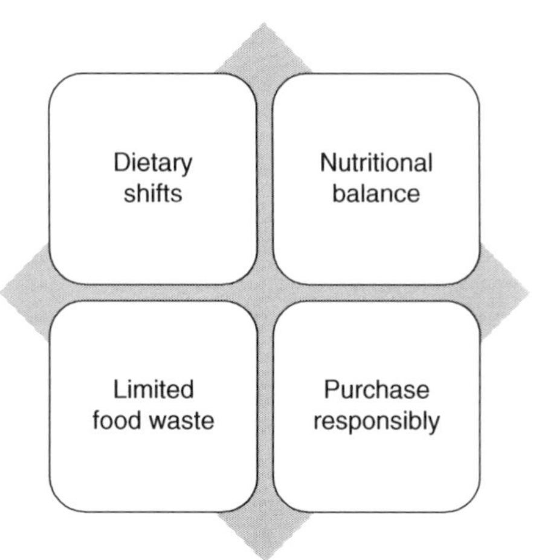


**Figure 9.1** Sustainable eating.

This is because consumers are the largest contributor to the food wastage carbon footprint, at 37% (FAO 2013). Industrialized countries are responsible for more waste from consumers than developing countries. This food waste is already 14% of municipal solid waste in the United States and it is rising (Canning et al. 2010; EPA 2012). All this waste demands landfill space and contributes to the habitat destruction associated with landfill expansion. Food waste comes with additional environmental costs when it degrades in the landfill by emitting methane, a GHG, contributing to climate change. Most importantly, wasted food means all the energy used for transportation, packaging materials, processing inputs, and agricultural production is also wasted.

Sustainable eating is comprised of four main elements: shifting diets from being heavy in animal-based and highly processed foods to more plant-based and fresh foods, not over or under-consuming calories and ensuring nutritional adequacy with a variety of foods, reducing wasted food, and purchasing responsibly-produced foods (see Figure 9.1). The United Kingdom is developing detailed principles on these points that include (DEFRA 2013):

- Eat a varied balanced diet to maintain a healthy body weight.
- Eat more plant-based foods, including at least five portions of fruit and vegetables per day.
- Value your food. Ask about where it comes from and how it is produced. Don't waste it.
- Moderate your meat consumption, and enjoy more peas, beans, nuts, and other sources of protein.
- Choose fish sourced from sustainable stocks. Seasonality and capture methods are important here too.

- Include milk and dairy products in your diet or seek out plant-based alternatives, including those that are fortified with additional vitamins and minerals.
- Drink tap water.
- Eat fewer foods high in fat, sugar and salt.

#### *9.1.1.1 Consumer behaviors*

Consumer actions of driving to the store and storing and preparing food at home also contribute to the food supply's environmental burden. In some cases, consumers are the dominant source of an environmental issue, above all other aspects of the supply chain.

Consumer transport to the store may sometimes outweigh all the impacts up to that point in the supply chain. This occurs when people travel to the store for the sole purpose of purchasing a few items and not conducting other errands on the same trip (Braschkat, Patyk, Quirin, & Reinhardt 2003). Consumers take about 122 trips to the grocery store per person per year (Coley, Winter, & Howard 2014). This adds up to over 360 miles (580 km) of driving for grocery shopping per person per year (Coley et al. 2014). One emerging approach that addresses this impact is home food delivery. Online food orders that get delivered to the consumer's home have a reduced impact to the environment (Coley et al. 2014). There is also an added benefit of reducing road congestion (Coley et al. 2014). There are other ways to reduce transport impacts from food shopping including combining trips, reducing the total number of trips, shopping at nearby locations, using more fuel-efficient modes of transportation, or using public transportation, a bicycle, or walking.

Household operations are the largest energy user across the supply chain (Canning et al. 2010). Household refrigerators and freezers require around 40% of total household food-related energy; cooking meals in stoves, ranges, and microwave ovens is around 20%; and heating water and operating dishwashers around 20% (Heller & Keoleian 2000). Outdated refrigeration is substantially less energy-efficient than current technology. There can be a two-to-four times difference in energy efficiency between ten-year-old and new equipment (James & James 2014). Shifting to newer and more energy-efficient appliances contributed to household food-related energy use declines of 15.7% overall and 19.5% per capita between 2002 and 2007 (Canning et al. 2010). During this period, energy prices increased 70%, encouraging the adoption of energy-saving practices (Canning et al. 2010).

### 9.1.2 Double pyramid

There is some correlation to the foods recommended to stay healthy and those that have a lower environmental impact. The Barilla Center for Food and Nutrition (BCFN) developed the Double Pyramid to illustrate that diets

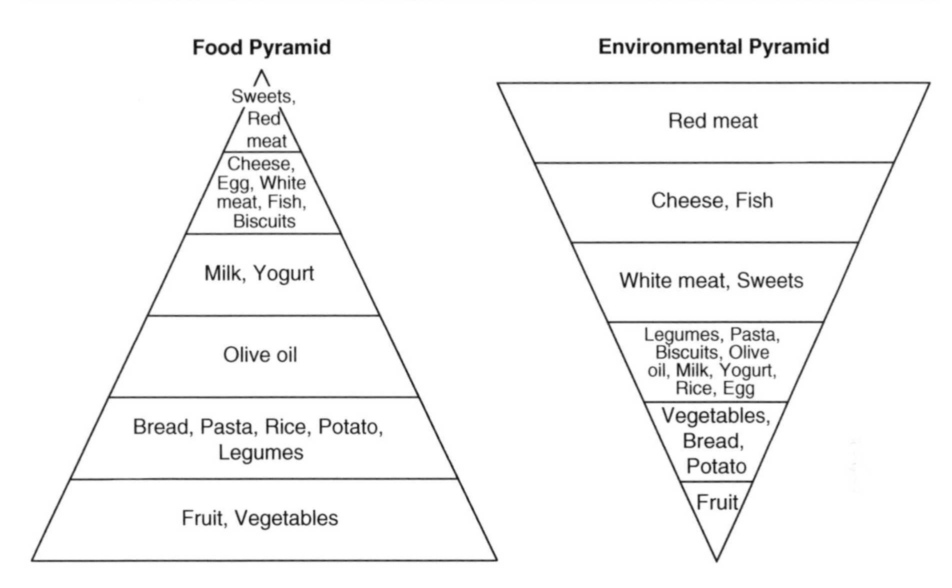


**Figure 9.2** Food and environmental pyramids (reproduced by permission of Barilla Center for Food & Nutrition Foundation - BCFN © BCFN 2013).

based on higher consumption of fruits, vegetables, and grains, and just a moderate consumption of animal products are both good for health and the environment (see Figure 9.2) (BCFN 2013).

Considering this concept, the Mediterranean diet is used as an example by the FAO and BCFN. They found that it is nearly aligned with the Double Pyramid based on pasta and other cereals, vegetables, fruits, and olive oil. Little red meat and limited fish and dairy is consumed. The Mediterranean diet is known to contribute to better health and well-being and has been found to have an environmental impact 60% lower than the North American-type of diet that typically has higher consumption of animal-based products (BCFN 2013; FAO 2012b).

## 9.1.3 Innocent promotes sustainable consumption

The United Kingdom food and beverage manufacturer, Innocent, has a corporate sustainability approach aimed to have a net benefit to the world. A key part of this effort is helping consumers understand their role in sustainability. The company provides consumers with information and recipe suggestions that align with sustainable diet recommendations. In addition, Innocent aims to get "the right balance between environmental impact and nutrition [to underpin] our entire sustainability strategy as well as being firmly embedded in our new product development process" (Innocent). The following is from the company's website (Innocent):

*Here at Innocent, we want to leave things a little bit better than we find them. A really important part of this is making sure that our products are good for you (full of healthy stuff and nothing weird) and good for the planet (produced with the least possible environmental or social impact). We call this sustainable nutrition.*

*What we eat every day is one of the main things that determines our impact on the planet. In the UK the food we eat (growing, producing and importing it) is responsible for 30% of our carbon emissions. Food can also have lots of other impacts including the cutting down of forests, heavy use of water for irrigation and pollution of water and soils from agrochemicals.*

*But here's the good news. A few small changes to what you eat each day can be really good for the planet (and really good for you too). Our friends at WWF UK have developed the five Livewell principles to help us make our food choices both healthy and green, and the best part is that there are no complicated rules to follow, no need to chop everything up into tiny pieces, and no need to start eating lettuce soup (unless of course you really like lettuce soup).*

*Five Steps for a healthier planet and you:*

- *Eat less meat*
- *Eat more plants*
- *Throw less away*
- *Eat less processed food*
- *Eat better food*

## 9.2 The potential of the principles

A sustainable food system is critically needed and requires action from all contributors in the supply chain. Producers, processors, distributors, marketers, and consumers play a role in implementing solutions that produce more food without compromising society or the environment on the same amount of land and reduce food demands, especially for resource-intensive foods (see Table 9.2). In short, produce more and consume better with a smaller impact (see Figure 9.3) (DEFRA 2013).

The case studies throughout the book highlight best practices that are already in place by some in the supply chain that are moving in the right direction, these are summarized in Table 9.3. With more progress, sustainability and feeding the growing population are achievable. But this requires change. The exciting part is that with change there is even the potential for the food system to provide an overall benefit to the economy, environment, and society by nourishing the population, revitalizing natural resources, enhancing economic

**Table 9.2** Mechanisms for implementing principles and practices to reach the potential

| | Farmers & Agribusiness | Manufacturing | Distribution & Marketing Channels | Consumers |
|---|---|---|---|---|
| **Take direct action (apply best practices)** | X | X | X | X |
| **Collaborate across supply chain** | X | X | X | |
| **Support farmer action** | | X | X | |
| **Change purchasing** | | X | X | X |
| **Educate and connect consumer to the value of sustainable production and supply chain** | X | X | X | |

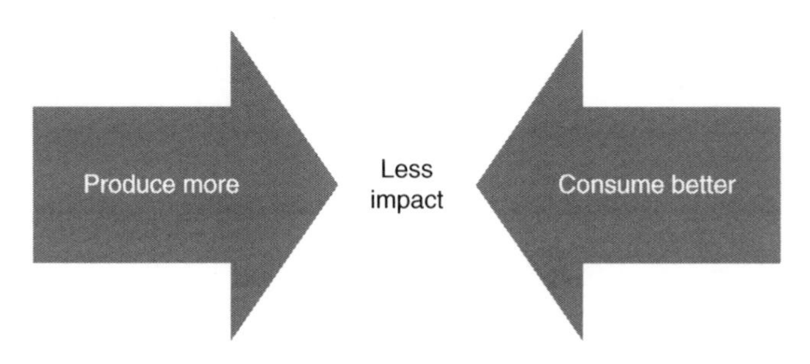


**Figure 9.3** Sustainable food system needs (DEFRA 2013).

development, and closing resource loops (Figure 9.4). This improvement can lead to a net good or net positive. Table 9.3 outlines how each Principle has this potential with greater adoption of the practices already available: the *principles–practices–potential of sustainability in the food industry*. Agricultural production can help replenish freshwater instead of being the dominant depleter of water. Animal rearing can restore soil and land rather than damage and degrade it. Food processing can generate renewable energy instead of demanding more energy from fossil fuels. Packaging can close the resource loop through recycling the resources instead of wasting them. The opportunities for each Principle have this type of positive potential. In addition, the examples throughout the book point to the business opportunities with each Principle including cost savings, reduced risks, growth opportunities, and improved employee satisfaction and productivity.

Progress with just one Principle is not sufficient, however. We need to achieve success with all of the Principles to claim true advancement toward sustainability. That is the key and why the Principles were drawn out in the book, to provide clear direction on where efforts need to be taken and progress

**Table 9.3** Ten principles of sustainability in the food industry summary with practices and potential: principles–practices–potential of sustainability in the food industry

| Practices | Potential |
|---|---|
| **1. Safe and highly nutritious food is accessible and affordable to promote and support a healthy population.** | |
| • Avoid microbial, chemical, and allergen contamination<br>• Reduce fat, sugar, and salt in processed foods<br>• Provide understandable and meaningful information about food to consumers<br>• Market food responsibly, especially to children and teens<br>• Ensure diverse, healthy, and nutritious food is available and affordable | • Wholesome food supply |
| **2. Agricultural production beneficially contributes to the environment while efficiently using natural resources and maintaining healthy climate, land, water, and biodiversity.** | |
| • Implement responsible production techniques such as:<br>  • Integrated pest management<br>  • Diverse crops and varieties with high yields<br>  • Crop rotations<br>  • Cover crops/green cover and no/low tillage<br>  • Managed irrigation<br>  • Soil management/nutrient management<br>  • Integrate conserved land or agroforestry<br>  • No artificially heated greenhouses<br>• Source and support responsibly produced crops | • Replenish water<br>• Revitalize biodiversity<br>• Build soil fertility<br>• Restore land and provide a carbon sink |
| **3. Use of animals, fish, and seafood in the food supply optimizes their well-being and adds to environmental health.** | |
| • Implement responsible animal and fish production techniques such as:<br>  • Well-managed grazing/feeding, land stewardship and stocking density, and effective manure management approaches (including responsible production techniques noted in no. 2)<br>  • Diverse breed production for health, resilience, and productivity<br>  • Productive rearing that respects the five freedoms of responsible animal care (freedom from hunger/thirst, discomfort, pain/injury/disease, inability to express normal behavior, and fear/distress)<br>  • Responsible seafood production and harvesting in the wild and in aquaculture (including responsible production techniques noted in no. 2)<br>• Source and support responsibly-produced and harvested food | • Replenish water<br>• Revitalize biodiversity<br>• Build soil fertility<br>• Restore land and provide a carbon sink |

**Table 9.3** Ten principles of sustainability in the food industry summary with practices and potential: principles–practices–potential of sustainability in the food industry (*continued*)

| Practices | Potential |
|---|---|
| **4. Producer equity and rural economy and development are strengthened with fair and responsible production and sourcing.** | |
| • Support smallholders and women in agriculture:<br>  • Ensure economic viability and share value<br>  • Build capacity<br>  • Improve infrastructure and rural services<br>• Source from local communities<br>• Integrate diversity of sources into procurement (with longer-term commitments for smallholders and women when possible) | • Enhanced economic development<br>• Resilient food supply<br>• Food security |
| **5. Safe and suitable working conditions are provided to support employees across the supply chain.** | |
| • Respect human rights<br>• Provide fair wages and labor conditions<br>• Empower workers<br>• Provide safe working conditions | • Build capacity<br>• Deliver safe and prosperous livelihoods |
| **6. Food and ingredient processing generates resources and requires minimal additional inputs and outputs.** | |
| • Conserve, reuse, and provide processing energy and water through lean, clean, and green processing:<br>  • Maintenance<br>  • Minor adjustments and process adjustments<br>  • Efficient technology, renewable energy, and water restoration | • Generate renewable energy<br>• Replenish water |
| **7. Packaging effectively protects food and supports the environment without damage and waste.** | |
| • Optimize each package's:<br>  • Functionality<br>  • Materials and sourcing<br>  • Processing and chemicals<br>  • Design and innovation<br>  • End of life<br>• Support recycling | • Close the resource loop through recycling<br>• Safe food supply |
| **8. Food and ingredient waste and loss are prevented across the supply chain and what cannot be avoided is put to a positive use.** | |
| • Prioritize food wastage management to:<br>  • Reduce<br>  • Reuse<br>  • Recycle/recover | • Feed hungry<br>• Reduce production needs<br>• Conserve resources |

(*continued*)

**Table 9.3** Ten principles of sustainability in the food industry summary with practices and potential: principles–practices–potential of sustainability in the food industry (*continued*)

| Practices | Potential |
|---|---|
| **9. Food and ingredients are efficiently delivered across the supply chain and to the consumer.** | |
| • Distribute efficiently and conserve fuel use through:<br>  • Lower GHG-emitting modes<br>  • Fuel-efficiency measures<br>  • Route changes<br>  • Alternative fuels<br>• Conserve, reuse, and provide processing energy and water through:<br>  • Maintenance<br>  • Minor adjustments and process adjustments<br>  • Efficient technology, renewable energy, and water restoration<br>• Reduce, reuse, and recycle waste<br>• Clean green and optimize indoor environmental quality<br>• Purchase environmentally preferable products and services<br>• Green construction and renovation<br>• Educate and connect consumer to the value of a sustainable food supply | • Generate renewable energy<br>• Replenish water<br>• Feed hungry<br>• Reduce production needs<br>• Close the resource loop<br>• Empower consumer sustainability |
| **10. The supply chain and consumers advance sustainable business and food consumption.** | |
| • Optimize consumer's:<br>  • Eating<br>    • Shift diets to more plant-based and fresh foods, and moderate animal-based and highly-processed food intake<br>    • Eat a balanced diet with a variety of foods to maintain a heathy body weight<br>    • Purchase responsibly-produced foods<br>  • Transportation<br>  • Storing and preparing<br>  • Food waste minimization | • Sustainable food consumption<br>• Support each of the other areas of potential |

achieved. Leading companies are already engaged in addressing more than one Principle as illustrated in the examples in the previous chapters:

- Unilever covered in crop and livestock production (Principles 2 and 3)
- Coca-Cola featured for processing, packaging, and distribution (Principles 6, 7, and 9).
- PepsiCo discussed in crop and packaging (Principles 2 and 7).
- Starbucks reviewed for crop, food security, and producer equity (Principles 1, 2, and 4).

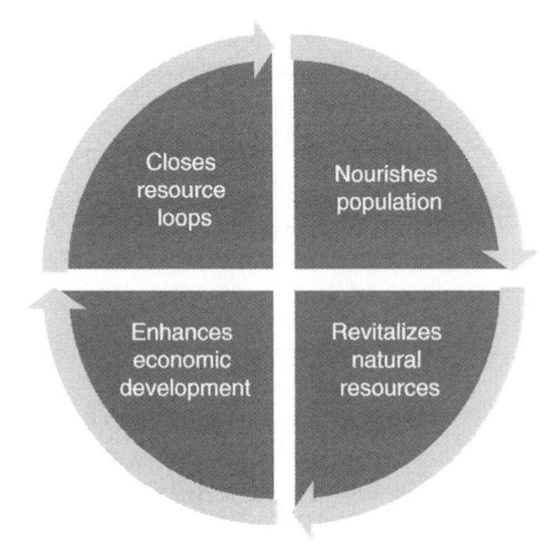


**Figure 9.4** Potential of the principles of food industry sustainability.

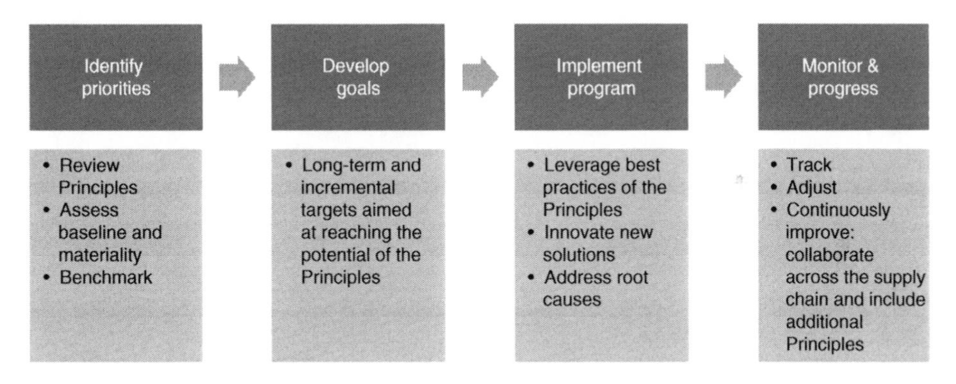


**Figure 9.5** Sustainability management process.

Each of these companies are involved in even more of the Principles, only parts of their sustainability programs were included in this book.

Importantly, the Principles can help programs form and develop (see Figure 9.5). Understandably, companies need to prioritize and focus their efforts to effectively advance progress and deliver business benefits (and ensure it is meeting the last principle of being a sustainable business, economically), but the Principles should be used to ensure that the relevant issues are carefully considered and to safeguard against unintended consequences or burden shifting. The Principles can also serve as a benchmarking tool to evaluate opportunities and progress. The potential of the Principles provides inspiration to position motivating goals aimed to deliver an overall benefit directly related to the Principles and proven programs. The best practices demonstrated through the supply chain should be adopted as well as new solutions developed to advance programs and address root causes. As efforts progress, additional Principles can be taken on in a similar way to reach the point of addressing all the material issues, much like the leaders are doing today. The opportunity to collaborate across the supply chain cannot be overlooked since it can enable more success on all of the Principles and provide innovation opportunities that may develop solutions faster.

Progress is being made and with more involvement from the supply chain the positive potential of the Ten Principles of Food Industry Sustainability can become a reality.

## References

Barilla Center for Food and Nutrition (BCFN). (2013). *Food and the environment: Diets that are healthy for people and the plant*. Retrieved from http://media3.barillagroup.com/cfn/magazine/BCFN_Magazine_FoodEnvironment.pdf

Braschkat, J., Patyk, A., Quirin, M., & Reinhardt, G. (2003). Life-cycle assessment of bread production—a comparison of eight different scenarios In *Life Cycle Assessment in the Agri-Food Sector: Proceedings from the 4th International conference*, October 6–8, Bygholm, Denmark. Retrieved from http://www.lcafood.dk/lca_conf/djfrapport_paper_2_poster.pdf

Canning, P., Charles, A., Huang, S., Polenske, K., & Waters, A. (2010). *Energy use in the U.S. food system*. Retrieved from http://web.mit.edu/dusp/dusp_extension_unsec/reports/polenske_ag_energy.pdf

CBC. (2013). *Food waste, overeating threaten global security*. Retrieved from http://www.huffingtonpost.ca/2013/11/23/food-waste-overeating_n_4328424.html?utm_hp_ref=fb&src=sp&comm_ref=false

Coley, D., Winter, M., & Howard, M. (2014). National and international food distribution: do food miles really matter? In B. Tiwari, T. Norton, and N. Holden (Eds.), *Sustainable Food Processing* (pp. 499–520), Ames, IA: Wiley-Blackwell.

Department for Environment, Food, and Rural Affairs for the United Kingdom (DEFRA). (2013). *Sustainable consumption report: Follow-up to the green food project,* Retrieved from https://www.gov.uk/government/uploads/system/uploads/attachment_data/file/229537/pb14010-green-food-project-sustainable-consumption.pdf

DeWeerdt, S. *Is local food better*? Retrieved from http://www.worldwatch.org/node/6064

Food and Agriculture Organization of the United Nations (FAO). (2013). *Food wastage footprint: Impacts on natural resources summary report*. Retrieved from http://www.fao.org/docrep/018/i3347e/i3347e.pdf

Food and Agriculture Organization of the United Nations (FAO). (2012a). *Improving food systems for sustainable diets in a green economy.* Retrieved from http://www.fao.org/fileadmin/templates/ags/docs/SFCP/WorkingPaper4.pdf

Food and Agriculture Organization of the United Nations (FAO). (2012b). *Towards the development of guidelines for improving the sustainability of diets and food consumption patterns in the Mediterranean area*. Retrieved from http://www.fao.org/docrep/016/ap101e/ap101e.pdf

Haley, M. *Changing consumer demand for meat: The U.S. example, 1970–2000*. Retrieved from http://www.google.com/url?sa=t&rct=j&q=&esrc=s&source=web&cd=1&ved=0CCwQFjAA&url=http%3A%2F%2Fwww.ers.usda.gov%2FersDownloadHandler.ashx%3Ffile%3D%2Fmedia%2F293605%2Fwrs011g_1_.pdf&ei=oZb2Us_jGqWEyAHYy4GQCg&usg=AFQjCNGNSzzhMjBpJlWnki-HT2_1Xz9BeA&sig2=oEaqiCsKktpJM01OYa6KFg&bvm=bv.60983673,d.aWc&cad=rja

Heller, M. C., & Keoleian, G. A. (2000). *Life cycle-based sustainability indicators for assessment of the U.S. food system.* Center for Sustainable Systems, University of Michigan, Report CSS00-04. Retrieved from http://css.snre.umich.edu/css_doc/CSS00-04.pdf

Innocent. *Being sustainable*. Retrieved from http://www.innocentdrinks.co.uk/us/being-sustainable/nutrition

James, S., & James, C. (2014). Sustainable cold chain. In B. K. Tiwari, T. Norton, & N. M. Holden (Eds.), *Sustainable food processing* (pp. 463–496). Ames, IA: Wiley-Blackwell.

Schneider, U. A., & Smith, P. (2009). Energy intensities and greenhouse gas emissions in global agriculture. *Energy Efficiency* 2, 195–206. doi 10.1023/A:1009728622410

Shelton Group. (2012). *EcoPulse '12: Overcoming the sustainability slump.* Knoxville, TN: Shelton Group.

U.S. Environmental Protection Agency (EPA). (2012). *Municipal solid waste generation, recycling, and disposal in the United States: Facts and figures 2012*. Retrieved from http://www.epa.gov/waste/nonhaz/municipal/pubs/2012_msw_fs.pdf

World Resources Institute (WRI). *Creating a sustainable food future.* Retrieved from http://www.wri.org/sites/default/files/WRI13_Report_4c_WRR_online.pdf

# Index



*The 10 Principles of Food Industry Sustainability*, First Edition. Cheryl J. Baldwin.

Printed and bound by CPI Group (UK) Ltd, Croydon, CR0 4YY